M.MALEKZADEH

SYSTEMATIC ERRORS IN ENGINEERING EXPERIMENTS

سوچهر ملک زاده
M.MALEKZADEH

SYSTEMATIC ERRORS IN ENGINEERING EXPERIMENTS

PETER POLAK

Department of Mechanical Engineering,
University of Sheffield

GULF PUBLISHING COMPANY
Book Publishing Division, Houston, Texas

This edition published
in the United States by
Gulf Publishing Company,
Houston, Texas.

ISBN 0—87201—819—9

Printed in Hong Kong

Published in Great Britain by
The Macmillan Press Ltd.

CONTENTS

1. TYPES OF ERROR   1

    Systematic versus random errors. Consistency, repeatability,
    acceptable errors in industry. Simple electrical examples,
    d.c. and a.c.

2. TEMPERATURE MEASUREMENT   6

    Thermal shunting errors. Thermocouples: laboratory and indus-
    trial set-ups, cold junction compensation, resistance correc-
    tion, new data on conduction losses. Compensating cables.
    Galvanometers versus potentiometers. Radiation problems.
    Kinetic heating effects. Mercury-in-glass thermometers, stem
    correction and zero errors. A common ice-pot error. Boiling-
    point error, superheat, barometric correction. Resistance and
    radiation methods. Melting indicators. Industrial recorders.

3. STATIC MECHANICAL PROPERTIES   23

    Machine extension versus gauge-length extension. Discussion
    of yielding in long and short specimens. Bending tests, local
    buckling in fibrous materials, friction errors, large deflec-
    tion errors, end errors in cantilevers, non-linearity in plas-
    tics, thermoelastic property of rubber.

4. DYNAMIC MECHANICAL PROPERTIES   31

    Stress gradient, grain size, residual stress due to prior load.
    Evidence from industry and tests. Constant load and constant
    deflection tests, relevance to real structures. Transfer of
    energy from machine to specimen. Chain-like and rope-like
    structures contrasted. Adiabatic impacts contrasted with
    likely loading in machinery. Inertia and overshoot.

5. FLOW MEASUREMENT   39

    Quantity and time methods. Timing errors, tank and float
    errors. Orifice and venturi meters, typical coefficients.
    Sources of error. Pulsating flow, viscous meter and errors.
    Pitot-static tube errors illustrated by photographs. Wall
    tapping error example. Hints on flow-visualisation, local
    accelerations of tracer particles.

PREFACE

This book is intended for students of mechanical, chemical and other
branches of engineering at universities or technical colleges. It
should help them to benefit from laboratory work by encouraging a
thoughtful, critical attitude and by indicating the wider context of
basic experiments. This type of material is difficult to present
and even more difficult to absorb when given in lecture form; it is
hoped that lecturers as well as students will welcome this selection.
It is complementary to the well-established books on instrumentation,
with a minimum of overlap.

The sequence has been arranged to suit students throughout their
tertiary education, starting with material which can be followed on
the basis of school-work, following on to those topics met early on
in an engineering course. The later parts include material appropriate to project and research work. Detailed descriptions have been
avoided, except in the case of gas analysis, in view of the limita-
tions of space and the great variety of apparatus available for most
of the other operations.

ACKNOWLEDGMENTS

The author is very grateful to Dr K. Martin of UWIST for his con-
structive criticism of the manuscript and to the University of
Sheffield for the use of general facilities. Particular thanks are
due to Mrs P. Windle, for typing and presentation, and Messrs
Hallford and Francis for help with photographic work.

SYMBOLS

E = Young's Modulus

ν = Poisson's Ratio

E' = $E/(1 - \nu^2)$

P = load (chapter 3)

N = normal contact force (chapters 3 and 7); otherwise newtons

μ = coefficient of friction (chapters 3 and 7)

θ = end slope (chapter 3)

σ = (tensile) stress (chapter 4)

k = concentration factor for stress or strain (chapter 4)

C_d = coefficient of discharge = actual flow ÷ theoretical flow at maximum measured or calculated velocity

Re_D = Reynolds number based on large (approach pipe) diameter = velocity × diameter ÷ kinematic viscosity

Re_d = ditto based on smallest (throat) diameter

g = acceleration due to Earth's gravity

Y = yield point (appendix 1)

I. TYPES OF ERROR

Laboratory experiments and practical measurements serve several purposes, some of which do not demand accuracy. Some experiments are exploratory in the sense of looking for the existence and direction of some effect before trying to establish its magnitude; others are chiefly instructional, to demonstrate theoretical principles. Some industrial measurements are needed only to control and repeat a process in accordance with previously established values, for example, heat-treatment of metal, food or textile processing, moulding plastics, foundry work. Systematic, consistent errors in such cases may be harmless.

When we have to communicate data, whether we are reporting research results or issuing instructions for operating a process on some new plant, the data cannot be divorced from the methods of measurement; as will be seen later, large errors can arise if the method of measurement is changed. In scientific work it used to be customary to describe apparatus in great detail since even if an experiment were carried out with the greatest competence there could still be alternative interpretations. With the great increase of published results, this custom has been dropped and it is no longer possible to study results in detail; if in doubt we have to attempt to repeat them.

Engineering students are chiefly introduced to the use of standard apparatus designed to demonstrate a particular effect; only the brightest students are likely to consider all the precautions which are (or are not) incorporated in the apparatus to minimise extraneous troubles. In industry we meet apparatus that faithfully records signals received from some sensor well hidden inside the plant and perhaps not disturbed for years. In the following chapters attention is drawn to the reasons why the signal from a sensor can depart from the true value it purports to measure.

There are several excellent books on instrumentation as used in industry and in the laboratory, also on statistical treatment of random errors. References to systematic errors are relatively brief. The student tends to assume that the importance of a topic is proportional to the area of print devoted to it; this is far from true. Those involved in experimental work personally and in enough detail to be familiar with its problems either do not find time to write books at all or tend to specialise; thus there are some very useful books each dealing with a limited range of problems. Here an attempt is made to cover a wide range of subjects relevant to the mechanical sciences and to manufacturing industries.

In the life sciences (including social sciences), natural variations between individuals and fluctuating natural conditions demand a statistical approach. In the mechanical sciences this is sometimes true also, for example, in fluid flow made turbulent by many sources of disturbance. In most cases, however, we have the situation under control, with uniform material able to give repeatable results. The test of repeatability is our first weapon; until we can repeat a result we must suspect that part of the system is not under control; the reason may be an obstruction, leakage, loose contacts, irregular draughts, unsteady supplies of electricity or fluid.

Repeatability is only the beginning. As will be shown later, temperature readings usually have negative error (that is, the reading is low). The magnitude of the error will vary with physical conditions such as pressure, velocity, the actual levels of temperature, time-dependent effects such as surface deposits, diffusion or corrosion. Flow rate readings also tend to be persistently low although there are some circumstances which give rise to positive error. Statistical methods do not claim to reveal this kind of error. In one sense it is not an error of reading but an error of interpretation; we have made a true observation not of the basic phenomenon but of the system phenomenon plus instrumentation.

A simple example should help to make this clearer. Suppose we wish to count the number of people using a park at various times and in various weathers, by installing a turnstile at one of the gates. The effects of this are obvious; many people will simply start using other gates, especially when the turnstile is wet after rain. So we install turnstiles at all gates. Immediately people with prams, children or dogs will tend to stop using the park if there are any alternatives. Thus we disturb the total and the distribution by the method of measurement; we do, however, get a true reading of actual users of the turnstiles, slightly distorted by children playing circular games or spinning the turnstile.

There are direct analogies to this in fluid flow measurement in the effects of blockage and swirl. The barrier effect also has some analogy with temperature measurement and with some forms of fatigue testing of metals. To sort out the true meaning usually demands some subsidiary tests.

Time-lag errors and friction errors can act in either direction depending on detailed procedure. With a regular experimental routine they may be consistent and easily overlooked. By intelligent use of reversals of loading, etc., the maximum error bandwidth can be found; the true reading is not necessarily midway between forward and reverse values because the situation is often unsymmetrical (see chapter 3).

The growth of modern instrumentation has made possible many measurements not previously feasible, such as continuous weighing on conveyor belts, various transient forces, vibrations or shock-waves but it has introduced new hazards. Many primary sensors give very small signals which have to be fed to a bridge or other amplifying system, then the output is measured and possibly digitised and typed

out by line-printer. At every connection there is a risk of a broken
coaxial cable centre-core, of a badly soldered joint or of loose or
oxidised contacts. The small primary signal may be interfered with
by magnetic fields from electric motors. Externally powered instruments may be affected by momentary drops in the power supply due to
load changes on the network. Some electronic controllers feed
strange wave-forms back into the mains which may introduce spurious
signals.

   Lest the student should get discouraged, let us conclude this
chapter by discussing some of the very simplest instrument effects,
amenable to simple correction calculations. Suppose we wish to
measure the d.c. resistance of some object X at various loads. The
circuit may be as in figure 1.1 or as in figure 1.2. In the first
case we measure the object current directly and also the voltage
drop across object and ammeter. In the other we measure the voltage
applied to the object directly and let the ammeter measure the sum
of object current and voltmeter current. The corrections are easily
made since the resistance of each instrument is stated and is substantially independent of load. (Be sure to try questions 1.1 and
1.2.).

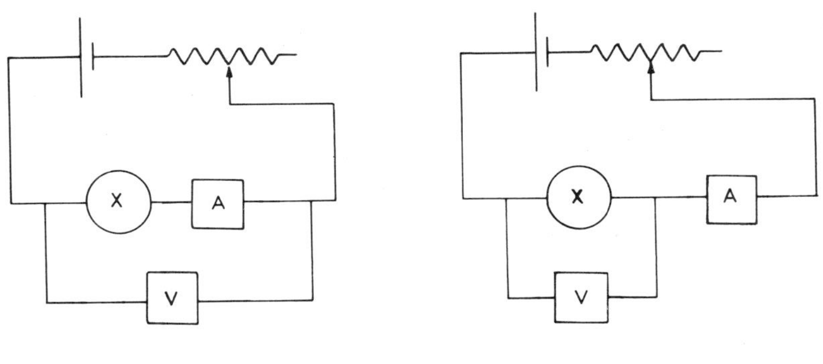

Figure 1.1   Figure 1.2

   The only source of confusion for students is the obscure way in
which some makers specify the resistance of multi-range voltmeters
or ammeters in ohms per volt or some similar expression. It is obvious that the resistance of the instrument is unlikely to be proportional to the voltage across it. The remark refers to the full-
scale value of the particular range. To make this clearer, consider
the circuit of a multi-range voltmeter or ammeter (figures 1.3 and
1.4). At full scale, the movement takes a certain current at a certain voltage drop. In the voltmeter case, the full-scale current is
the same on all ranges, the voltage drop being the sum of the drops
in the movement and in the added series resistances. Conversely in
the ammeter case the voltage drop at full scale is common to all
ranges and the current is divided between the movement and the appropriate shunt.

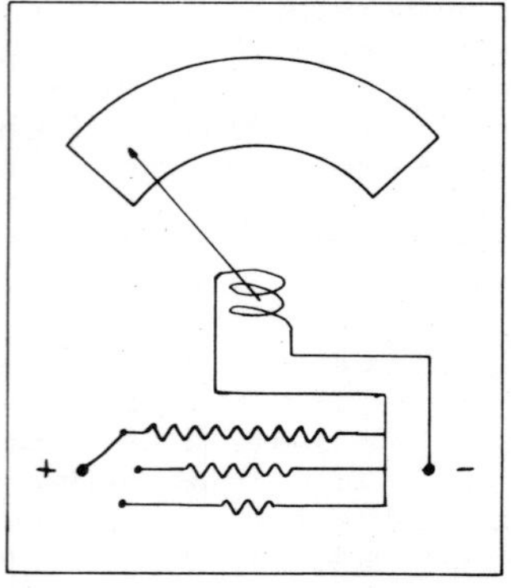
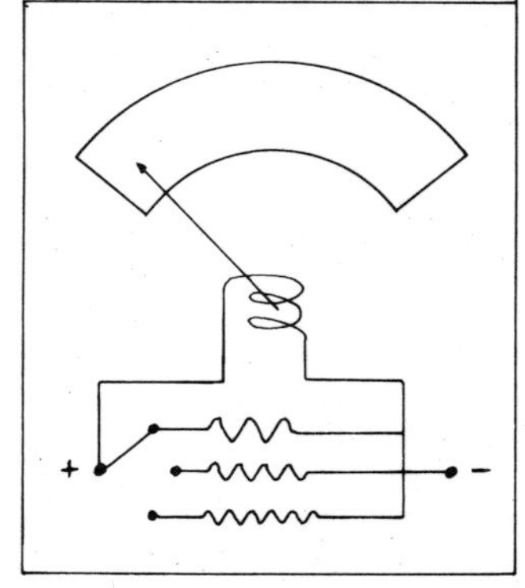

Figure 1.3   Figure 1.4

In the case of alternating current the situation is complicated
by phase changes. The meter will have a very slight amount of inductance; the object (load) may be predominantly inductive or capacita-
tive. It is assumed that the student has studied simple a.c. theory
and should be able to solve such a circuit problem. As an illustration, an example is shown of a traditional method for measuring reac-
tive circuits. It requires adding a pure resistance R to the circuit
(figure 1.5). If the value of this resistance is known at all values

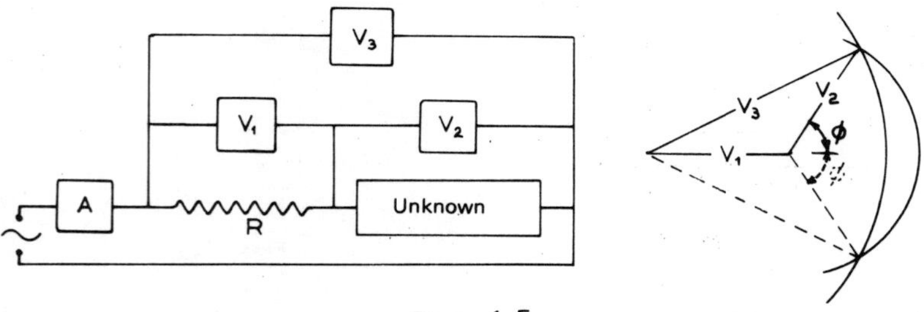

Figure 1.5

of current, the ammeter can be omitted since the current can be calculated from voltage V_1. The voltage across the whole circuit is
less than the sum of the individual voltages; the ratio $V_3/(V_1 + V_2)$
gives the power factor for the circuit. By constructing the vector
diagram we can also find the phase angle and power factor for the unknown load, although we cannot by this simple system distinguish between lead and lag. It is usual to assume that the currents taken by
the voltmeters are negligible, but a correction is easily applied.

4

QUESTIONS

1.1   If the circuit of figure 1.1 uses an ammeter of 1 mΩ resistance
and a voltmeter of 1000 Ω resistance, find the current and voltage
drop in each part of the circuit when the current through the load X
is some convenient value, the load having a resistance of (a) 2000 Ω,
(b) 1500 $\mu\Omega$. What resistance value would have been obtained in each
case by considering the voltmeter and ammeter readings only, without
correction?

1.2   Repeat question 1.1 using the circuit of figure 1.2. Now medi-
tate on the suitability of each arrangement for the given measuring
tasks.

1.3   Let figure 1.3 represent a multi-range voltmeter whose movement
has a 10 Ω resistance. The movement gives full-scale deflection
when carrying a current of 500 μA. Find the resistance which must
be put in series to result in the following ranges: (a) 1000 V,
(b) 250 V, (c) 1 V, (d) 10 mV, (e) 5 mV.

1.4   Let figure 1.4 represent a multi-range ammeter using the same
movement as in question 1.3. What shunt values are required to produce the following ranges: (a) 50 A, (b) 1 A, (c) 5 mA (d) 0.5 mA?

1.5   You are told to measure the growth rate of a field of lettuces
by cutting and weighing 1 per cent of the original crop at weekly
intervals. What influence may the removal of the samples have on
the rest of the crop? Suggest some sampling ideas to minimise these
influences. What time of day would you consider likely to give the
most consistent results?

2. TEMPERATURE MEASUREMENT

Perhaps the most serious and least suspected errors arise in temperature measurement. It may be unfortunate that students are introduced to the measurement of water temperatures before any other; water is an excellent heat-transfer medium so that the sensor is in good thermal contact, whereas large errors can arise in measuring the temperature of gases or poorly conducting solids.

The difficulties may be broken down as follows.

(a) Recognising how much the sensor and its access port disturb the system.

(b) Deciding whether kinetic and radiant heating corrections are required.

(c) Finding the necessary corrections to the sensor output, sometimes by a subsidiary experiment.

(d) Assessing whether the reading has only local significance or represents a larger region.

2.1 FLUID TEMPERATURES

If we immerse a thermometer or other sensor directly in the substance under test, there will be an interchange of energy until equilibrium is established, where the heat flow into the bulb by convection, conduction and radiation equals the heat leakage outwards by conduction along the stem and by radiation across the fluid to distant surfaces. In a glass thermometer the stem loss is generally very small; along thermocouple wires it can be large. Radiation between sensor and walls can be serious in hot gases and is discussed in section 2.5.

It is often convenient to confine the fluid so that sensors can be changed without shutting down the system; the method used is a pocket or thermo-well. At high pressures the use of a pocket also avoids errors with glass thermometers due to elastic compression of the thin glass bulbs which would squeeze up some mercury and give a false reading. The sensor will tend to read the temperature in the pocket, especially if it is filled with a little oil to improve thermal contact and if air circulation is prevented. In water, the pocket may be essentially at water temperature but in gas large errors can arise.

For example, an airflow at 10 m/s (which is quite a fair velocity for heat-transfer purposes) can give the errors shown in figures 2.1a and b, where the temperature within the well is influenced by heat leakage along the well and to the atmosphere. A way of diminish-

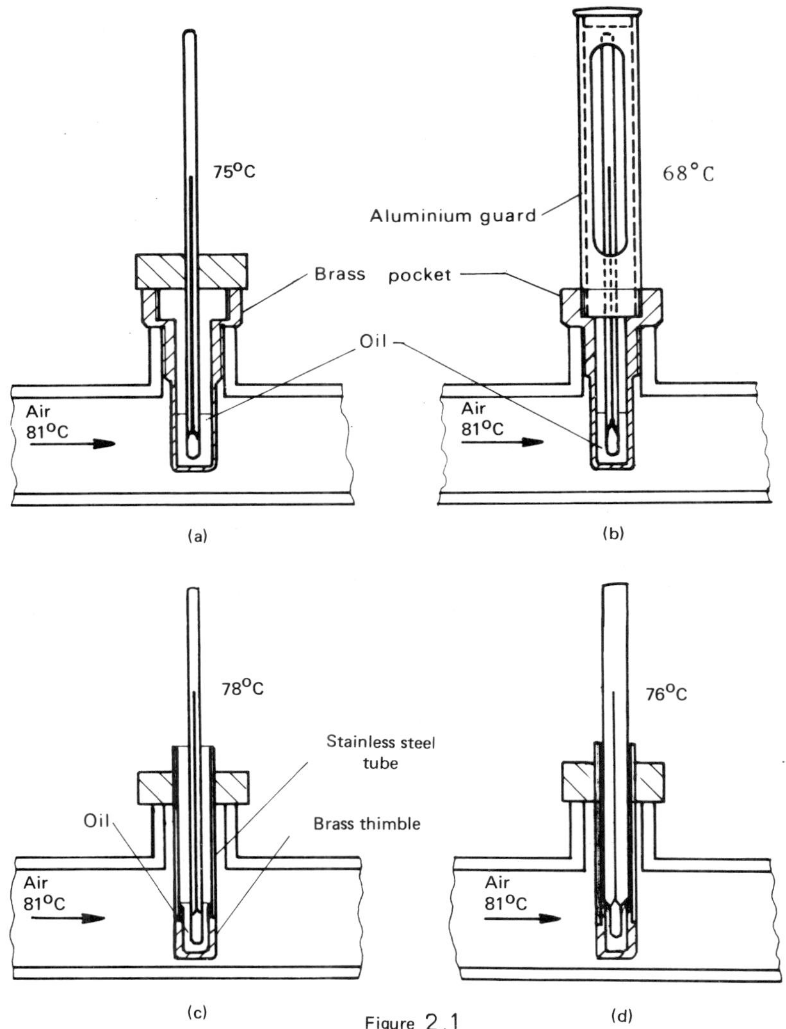

Figure 2.1

ing the error is to make the well bulb from a good heat conductor and
the neck from a poor heat conductor, for example, brass and stainless
steel, figure 2.1c. This, however, can prove illusory, if the thermometer bulb does not fit the base of the pocket. Figure 2.1d shows
what can happen if the thermometer is too fat in the stem. See also
section 2.6.

2.2   THERMOCOUPLE PROBLEMS

Thermocouples are so widely used that their construction, use and
problems are discussed first, also touching on aspects general to all
types of sensor.

In a circuit consisting of two different conductors, raising the temperature of one junction above that of the other produces an electromotive force (e.m.f.) which varies progressively with temperature difference though not necessarily following a linear law. Almost any pair of metals could be used; to avoid confusion, general usage is confined to a small range of materials. Their exact composition and the e.m.f. to be expected are laid down by various standardising bodies; therefore they can be referred to by their descriptive names knowing full well that these names are backed up by precise definitions. The usual pairs are copper/constantan for -250 to +400 °C, iron/constantan for -200 to +850 °C, Ni-Cr/Ni-Al, often called by trade names such as Chromel/Alumel or T_1/T_2, for -200 to +1100 °C. Some high-temperature couples use tungsten and other metals. All the preceding are classed as base-metal couples. In addition, there are six recognised rare-metal couples based mainly on platinum. The student will generally meet only the first three base-metal couples and Pt/Pt-10%Rh.

The uniformity of composition maintained by manufacturers of thermocouple wire is supposed to be such that any couple may be expected to give an output within about 1 per cent of tabulated standard values for the commoner base-metal couples. If greater accuracy is needed, individual calibrations can be made provided that an accurate master is available and installation errors are avoided. Rare-metal couples are more accurate; BS 1041 shows accuracies of ±1 °C for Pt/Pt-Rh over the range 0 to 1100 °C, ±2 °C from 1100 to 1400 °C. In many cases the errors arising from installation will be *considerably more* than this.

Thermocouple wires can be joined in various ways: gas welding in a small oxy-acetylene flame, resistance spot-welding, silver-soldering, etc. The presence of a third metal does not affect the output provided that it all stays within the hot-junction zone; the effective hot-junction temperature is where the uncontaminated thermocouple wires emerge. Soldering weakens very fine wires so that they need good support; sometimes the wires are twisted together before the junction is formed. The oxide film is usually strong enough to prevent shorting; any shorting within the zone which is at hot-junction temperature will not affect the output.

After prolonged use a thermocouple may have changed its composition, by diffusion of the two metals into each other, diffusion of the solder or brazing alloy, absorption of furnace gases, or oxidation. It is wise to recalibrate or replace old thermocouples.

Tensile stress and magnetic fields may have an effect on output. For instance an iron/copper couple with junctions at 0 and 100 °C at 0.01 tesla field and no stress has 1.5 per cent excess output; at 200 N/mm^2 tensile stress and 0.08 tesla field its output is 1.4 per cent low. [1] The author is not aware of similar data for practical couples (iron/copper is unusual owing to an inversion in its e.m.f. above about 500 °C).

In the laboratory one can use the layout of figure 2.2a which is easy to understand and allows the use of e.m.f. tables directly since these are normally given for a cold (or reference) junction at 0 °C. Note the possible error described in section 2.6. So long

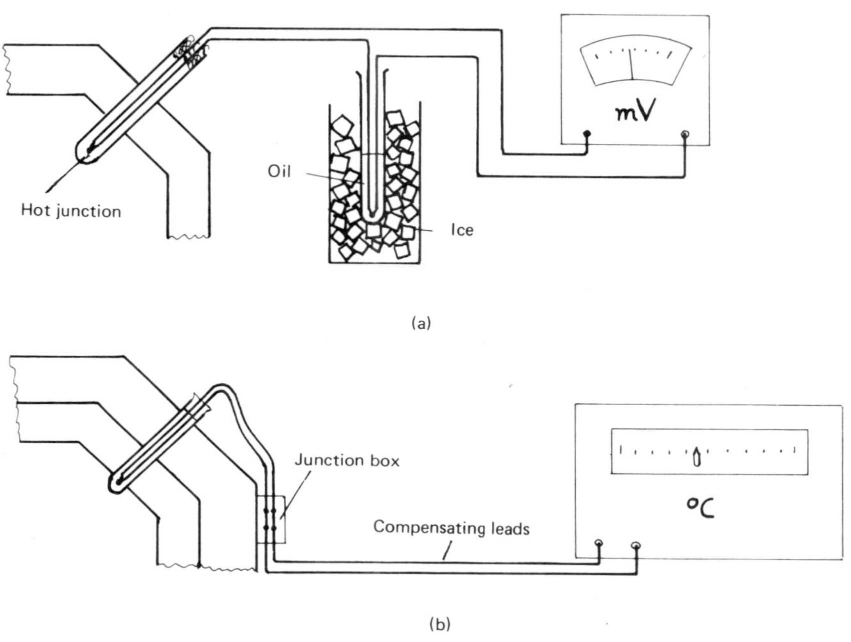

Figure 2.2

as both terminals of the instrument are at the same temperature, no e.m.f. error should be produced there.

An industrial set-up is more likely to be as in figure 2.2b. This makes the effective cold junction at the instrument terminals. Many industrial instruments have automatic cold-junction compensation (see section 2.11).

A galvanometer-type instrument takes an appreciable current; its reading refers to the voltage across the terminals, which is less than the e.m.f. round the circuit, therefore a resistance correction is required if not already incorporated in the scale. The situation is best explained by assuming we have a voltmeter of known resistance R_m, the external thermocouple circuit having a resistance R_c. If the true e.m.f. = E, the current is $E/(R_m + R_c)$. The instrument reads the voltage across the meter terminals, that is, current times

R_m; thus the true e.m.f. is current times total resistance and is obtained by multiplying the reading by $(R_m + R_c)/R_m$ (this is analogous to the internal resistance effect in batteries). Owing to the failings of human arithmetic, students are often advised to use a potentiometer. A potentiometer uses a null method and takes no current from the couple once the point of balance is reached, so that no correction is needed for resistance. However, a manual potentiometer has several disadvantages over a voltmeter. It is more expensive, slower and the battery gets tired and needs to be restandardised; the chief disadvantage is, however, that the reading is intermittent, so that oscillations of the signal are not detected. The momentary test current disturbs the system. Bad contacts or intermittent short circuits are not detected.

Digital-display electronic voltmeters generally take very little load from the system, being of high impedance.

Battery or mains-powered electronic voltmeters with a galvanometer output are convenient and foolproof but can impose a false reading on the system since the internally generated voltage is applied to the system cyclically in order to detect the point of balance. The author has personal experience of obtaining a reading 10 per cent higher than by several orthodox methods, on a differential temperature measurement. It is believed that the instrument caused a forced current flow through the circuit which used very fine wires, causing Peltier effect cooling of the cold junction.

Instruments using alternating current are slightly subject to spurious effects due to magnetic fields, self-inductance and capacitance. It is important to avoid long lengths of coiled-up wires; leads should be parallel and fairly close together to enclose a minimum of electromagnetic field.

2.2.3 *Installation*

Thermocouples do not last forever, they suffer oxidation, corrosion, thermal fatigue or contamination of the wires as mentioned above. For convenience of replacement and also to reduce cost and resistance, a thermocouple is often made quite short and connected to the instrument by compensating leads. If copper leads were used, the effective cold junction would be at the junction between thermocouple and copper wires. If this is close to the apparatus, it may well be at an unknown temperature. Compensating leads are made to have the same e.m.f. as the couple at least over the temperature range likely to occur at the connection. Some are made of slightly off-standard thermocouple wire, of heavy gauge to keep the resistance low, others of specially blended alloys. In effect they extend the thermocouple right up to the instrument.

When checking the resistance at operating temperature, the e.m.f. interferes slightly with the measurement. It is wise to measure the resistance in both directions and take the mean. Do *not* measure the instrument resistance, you may easily overload it. Only measure the *external* circuit resistance.

When measuring water temperature it is very tempting to immerse
the couple directly in the water, gaining good thermal contact.
There are, however, two dangers: the creation of electrolytic
e.m.f. of the order of 500 mV or more while the thermal e.m.f. is
only a few millivolts, and a minor error by electrical leakage
through the water.

2.3 CONDUCTION ERRORS IN THERMOCOUPLES

It is not generally realised that conduction along thermocouple
wires can cause very large differences between the observed temperature and the temperature intended to be observed.

Figure 2.3 shows the percentage error in readings given by several
different thermocouples. This includes radiation error but the greatest part is due to conduction along the wires to cooler parts. Fine
wires give less error, Chromel/Alumel gives less conduction error
than copper/constantan of the same wire diameter. Note the curling-
back method of increasing the effective amount of immersion.

The experiments on which figure 2.3 is based were carried out at
various temperatures ranging from 60 to 150 °C. At these temperatures
radiation errors would generally be small compared with conduction
errors. The heat flow from air to thermocouple junction and from the
wires to the surroundings will be essentially linear with temperature,
justifying the form of presentation on a per cent error basis.

The errors stated are entirely thermal; any resistance corrections
have been taken care of. This produces the interesting point that
when thoughtlessly using a voltmeter one could obtain the same crude
reading from two different couples, one showing a large thermal error
plus a small electrical error, the other a large electrical error
(fine wires) plus a small thermal error. It is evident that this is
mere coincidence, the errors happening to add up equally.

Fine wires are fragile and may easily be bent out of position by
the fluid. It is usual to enclose them, either in a thin stainless
steel sheath or in a refractory sheath. Examples of the errors due
to sheathing are included in figure 2.3.

Some commercial thermocouples are welded to the metal sheath near
the tip; these minimise the conduction error since in effect they
have an enlarged heat-collecting area. They have a disadvantage in
that whereas a loose couple in a sheath can be replaced at any time,
a fixed couple cannot be readily removed while the plant is under
pressure.

2.4 RADIATION ERRORS

If we measure temperatures in a uniform field such as in a furnace or
gas flowing in a well-insulated pipe, we may expect radiation between
sensor and walls to be small. In liquids which generally are not
very transparent to radiant heat, again the errors will be small. If,
however, we measure hot gas in a cool pipe, or in free air, or a cold
airstream in a hot furnace, the radiation to or from the sensor can

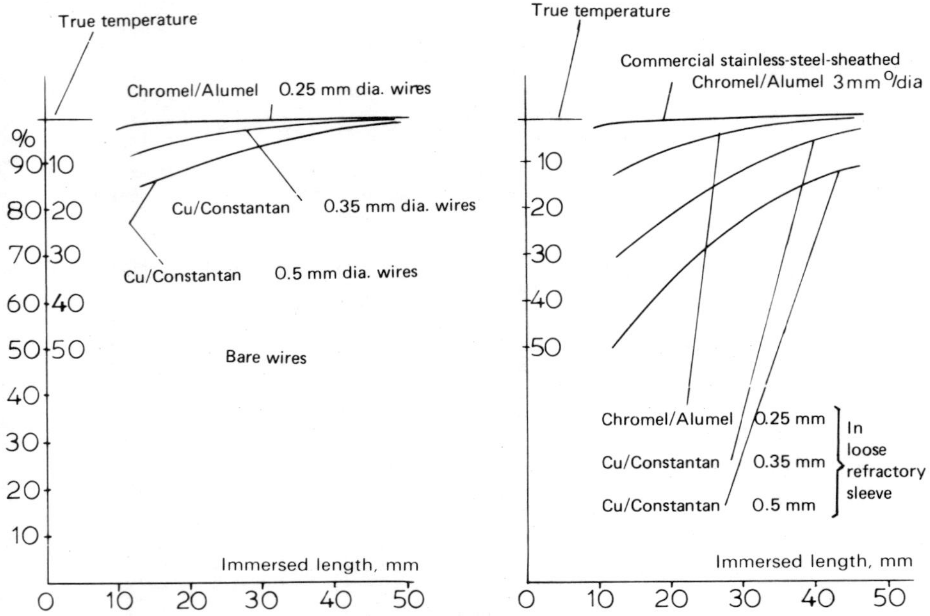

Errors at airspeeds ∼ 2.5 m/s

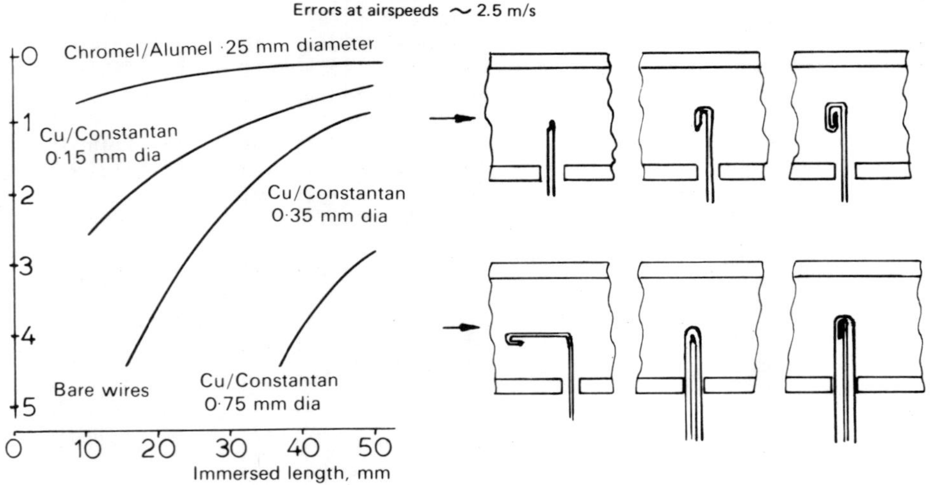

Errors at airspeeds ∼ 10 m/s

Figure 2·3

be substantial and errors of hundreds of degrees can arise, because
although the sensor has convective contact with the gas, the air or
gas readily transmits radiation so that the sensor also responds to
the temperature of the more distant surroundings. [2]

This can also affect furnace temperature readings just after
loading the furnace with a cool body; the reverse effect could
occur in a pipe coated internally with glowing carbon deposits.

Commercial thermocouples can be obtained with radiation shields
which take up a temperature that is generally fairly close to the
gas temperature. If active combustion is present, such shields can
have a second effect, disturbing the flame and thus giving misleading local conditions.

One of the most systematic methods is the use of the suction pyrometer, in which a sample of gas is drawn over a thermocouple at a
speed controlled by the operator. The higher the suction rate, the
stronger is the convective heat transfer so that when a further increase of suction rate gives little or no further change, a fair
reading has been obtained. This method is, however, open to the
objection mentioned above if there is any combustion going on; the
high local speed can either enhance or depress the combustion rate.
The temperature reading undoubtedly has a physical (or chemical)
meaning but it may not be the meaning we wish to capture.

In short, the interpretation of high-temperature readings is a
major topic beyond the scope of ordinary undergraduate experiments.
This is not to say that high-temperature measurements are invalid;
most industrial measurements are made for control purposes to ensure
that a process is being operated in a way which has previously been
found satisfactory. The difficulty will only become serious when a
new plant is being set up following instructions to operate at a
certain 'temperature' while the method of observing this temperature
is not fully specified. Quite apart from any radiation shielding,
the shape, layout, surface emissivity and finish of the sensor can
affect the balance between convective and radiant heat-flow.

Combining the difficulties of radiation and conduction errors, it
becomes evident that measuring a temperature profile in a non-uniform
field is likely to give a result depending as much on the method as
on the basic phenomena; results should be viewed accordingly.

2.5 KINETIC EFFECTS

The temperature of a solid expresses the mean molecular vibrational
energy at the point in question; in a fluid the energy due to
directed motion may be significant, making it necessary to distinguish between static and stagnation temperatures. The static temperature, t, relates only to the random molecular motion and it is this
temperature which is referred to for determining the fluid's properties such as density or viscosity. If, however, the fluid meets a
body with high relative speed, the kinetic energy will be wholly or
partly converted into random motion (heat). It is this effect which
produces kinetic heating of aircraft and missiles and which causes

most meteorites and other fragments from space to vaporise and
oxidise safely before hitting the Earth.

When measuring temperature, this effect occurs also at the surface
of the temperature sensor. The highest temperature is reached on the
leading surface of a body where the velocity is fully arrested. Here
the fluid reaches the stagnation temperature $T = t + u^2/(2c_p)$ where u
is the fluid's directed velocity and c_p is the specific heat measured
at constant pressure and expressed in relevant units. At the sides
of the body the fluid is not arrested to the same degree, reaching
a temperature somewhere between t and T. On the downstream face
conditions are uncertain, especially at supersonic velocities when
the mean molecular velocity and the flow velocity become comparable;
for example in nitrogen at 0 °C the mean molecular velocity is 454
m/s, the speed of sound being 337 m/s. [3] At supersonic speeds the
downstream face of a body has relatively little thermal (molecular)
contact with the fluid. A thermocouple tip pointing into a fast
fluid stream may come close to the stagnation temperature since heat
transfer will be most vigorous near the front, subject of course to
radiation problems discussed in section 2.4 and conduction along the
wires.

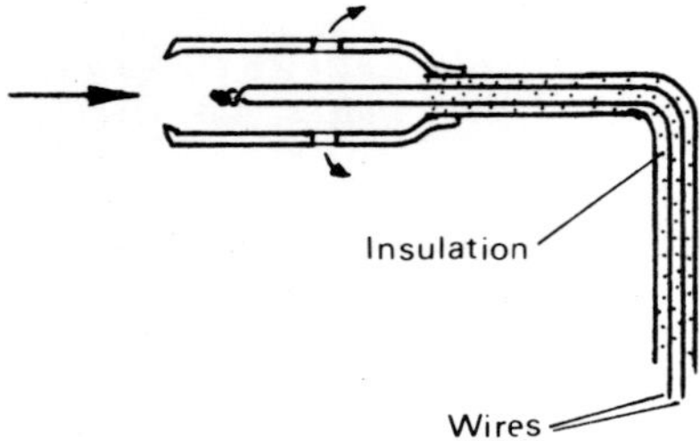

Figure 2.4

When the actual significance of the temperature is particularly
important, as for instance when it is used to estimate losses in
water turbines or pumps, the sensor can be made small and surrounded
by a stagnation pocket as in figure 2.4. A small throughflow is
desirable to ensure that the fluid within the pocket is a fresh, up-
to-date sample. In such a situation it is particularly important to
sample the whole stream cross-section carefully since the losses may
be very unevenly distributed in the outflow.

2.6 MERCURY-IN-GLASS THERMOMETERS

Domestic thermometers can be filled with alcohol, pentane or other
fluids but thermometers met in the laboratory normally use mercury,
with nitrogen gas above the mercury. Unless specially certified by
a Standards authority, temperature intervals may generally be expected to be accurate to 0.5 per cent of full range. Glass thermometers can acquire substantial zero errors, either if several years
old or if in continuous service at high temperature. The reason for
this is that glass is a supercooled liquid and flows (creeps) slowly
under stress.

In accurate work, expansion of the mercury in the stem as well as
in the bulb needs considering. Thermometers are graduated either
under fully immersed conditions or with a standard immersed depth of
50, 76 or 100 mm, indicated by a small circular mark on the stem.
Often we have to use a thermometer under conditions differing from
the calibration condition, making a correction by calculation. The
length of stem above the mercury is unimportant; the mercury thread
has +1 per cent error for every 60 °C of excess temperature. [4] To
make this clearer, consider figure 2.5. The vessel shown has room
for three thermometers. At (a) we have a thermometer calibrated for
full immersion, reading 80 °C and immersed up to the 30 °C mark
which comes midway inside the stopper. The ambient temperature is
20 °C so that the exposed 50° length is too *cold* by 60 °C and is
therefore in error by -1% × 50°, that is, -0.5 °C. The true temperature is therefore 80.5 °C. Thermometer (b), situated above the

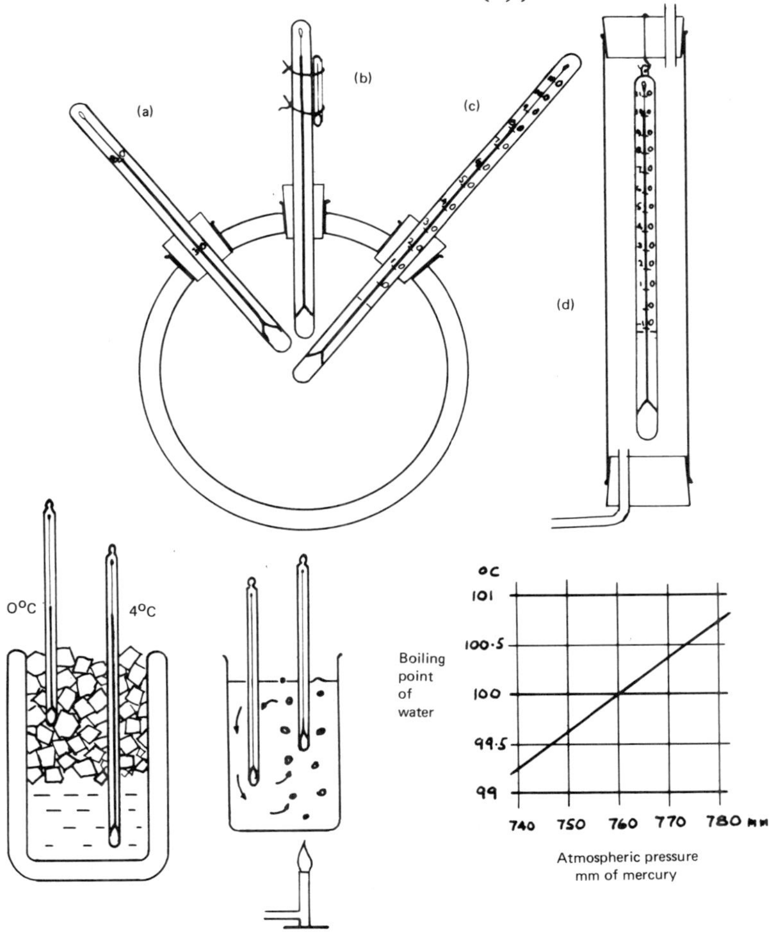

Figure 2.5

15

apparatus, is not necessarily at ambient temperature; the relevant
stem temperature could be estimated by securing another thermometer
to it as shown. At (c) we have a thermometer calibrated for 50 mm
immersion but immersed much further than this. The immersed excess
is equivalent to 30 °C and is too hot by 60 °C, having been calibrated at 20 °C stem temperature; it is therefore in error by
+0.3 °C (that is, 1% of 30 °C). Finally at (d) we have a thermometer calibrated at 76 mm immersion hanging in a steam bath at 100 °C.
This particular one has the immersion mark just below -10 °C so that
we have an effective column 110 °C long and too hot by 80 °C. ˙·es
This gives an error of (80/60)% of 110 or 1.4 °C.

An odd feature can mislead the student who tries to check a
thermometer at 0 °C. A convenient set-up is a Dewar flask (thermos-
jar) filled with ice. Fresh ice from a refrigerator is well below
0 °C; the surface of the melting lumps is no doubt at 0 °C but a
thermometer with its bulb in the bottom of the jar can read up to
4 °C. The reason seems to be that at 4 °C water attains maximum
density; any water not at 4 °C rises to the top, as does the ice.
The thermometer needs to be placed among the lumps of ice but clear
of any accumulated water. The same trouble occurs when an ice-pot
is used as the cold junction in a thermocouple circuit.

Testing a thermometer at 100 °C is subject to a few minor problems: onset of boiling can be obscured by the evolution of dis-
solved air well before boiling point is reached; very pure, air-
free water can superheat considerably before boiling starts; uneven,
localised heating gives uneven temperature in a vessel; finally,
the boiling point of water is affected by the local barometric
pressure (readers of Mark Twain will recall the use of this principle
to find altitude). These points are brought out in figure 2.5. Note
the form of presentation adopted: the grid lines are extended to
emphasise visually that the base is *not* at zero.

A safer way of testing thermometers around 100 °C is to place them
in the steam above boiling water so that it is in equilibrium with
condensing steam rather than the more dubious boiling water.

2.7 RESISTANCE DEVICES

The student is likely to meet two kinds: the platinum type and the
thermistor. Platinum resistance thermometers are used as basic
standards because of their high, repeatable purity, resistance to
oxidation and the great accuracy with which electrical resistance
can be measured. This tends to cloak them with a magic aura.

The element is usually enclosed in a protective casing to protect
the platinum from knocks, and from 'poisoning'. Platinum at high
temperatures reacts with hydrogen and some other substances (hence
its use as a catalyst); it is liable to be sufficiently affected
to alter its resistance slightly to an uncertain extent; this is
known as poisoning. A consequence of the enclosure is that thermal
contact with the surroundings is hindered. Installation-dependent
errors are still possible, from the enclosure leaking heat direct to
the supporting stem.

When connecting a platinum resistance element to its instrument,
the resistance of the leads can be compensated by using dummy leads,
wired into the bridge circuit in such a way that only the element
resistance changes are detected. Unfortunately, electrical connections are necessary since it would be too expensive and fragile to
take the platinum wires in one piece right up to the instrument.
The junctions between element and extension lead may corrode, oxidise
or loosen in time, which is more serious than with thermocouples
since a small change in contact resistance may be large when compared
with the signal, which itself is a small resistance change, whereas
in thermocouple circuits the total resistance may be hundreds of ohms.

Another possible source of error is thermocouple e.m.f. at the
junctions due to differences of temperature and due to the currents
flowing through the bridge. The measuring currents are kept very
small; a.c. is often used; in the vicinity of electric furnaces,
etc., the cables need to be screened.

Thermistors are an extremely useful class of device in the laboratory. They are small sensors, based on semiconductor materials, and
exhibit large resistance changes with temperature, positive or negative, over a narrow range. A wide choice is available, each with a
predicted calibration. The chief precaution to be observed with
these is to avoid overheating. In measuring resistance, an electric
current must be supplied, at least until the bridge is balanced.
The small size of a thermistor limits its heat-dissipation power.
If it is accidentally supplied with more electrical power than
specified, *especially if it is already at elevated temperature,* it
can remain operative but change its calibration. If used in important experiments, the calibration should be checked before and after
to confirm that this has not occurred.

A thermistor bead can be nearly as small as the proverbial bee's
knee. A bare bead of 0.6 mm diameter with two 0.1 mm wires is a
standard catalogue item; glass-covered beads are slightly larger.

Further details should be obtained from manufacturers' literature.

2.8 RADIATION PYROMETERS

For furnace temperatures, pyrometers based on radiation are often
used. These consist of a telescope with special lenses of material
transparent to a broad spectrum of wavelengths. The temperature in
the field of view is inferred from either the colour and brightness
when matched against an electrically heated filament previously
calibrated for temperature versus current, or by focusing the rays
on to a target whose temperature is measured by thermocouple, or by
comparing the energy contents of two separate parts of the spectrum.

The most obvious error sources are smoky atmospheres in the
furnace room, a fine deposit of dirt on the optics, or bad sighting
so that the view includes some parts not at full temperature. The
more elaborate instruments have a feed of air to keep the windows
clean in use, provided they are cleaned before use.

Less obvious are the fundamental difficulties of interpreting

the reading. An empty furnace corresponds closely to a black body
since any entering radiation is most unlikely to be reflected provided the peephole or furnace door is small. The interior gases
and walls may well be in equilibrium, though it should be borne in
mind that CO_2 and H_2O vapour have appreciable absorptivities and
scattering power at some wavelengths. We generally avoid sighting
on to electric elements or burners. The difficulty gets more
serious when an object is placed in the furnace. Bright metals or
smooth refractories will reflect radiation from the furnace walls
and will immediately appear to be almost at furnace temperature
whatever their actual temperature. Fortunately many objects are
not highly reflective but have high absorption at typical furnace
temperatures, but unfortunately the absorptivity and emissivity
vary with body temperature and wavelength so that the radiation
given out is a mixed value. This is an instance where satisfactory
repetitive results for industrial purposes can be obtained without
the finer points of science so long as the methods of measurement
do not change.

2.9   TEMPERATURE IN SOLIDS

The temperature in a solid is not measurable without some disturbance
of the heat-paths. The least disturbance is likely to come from a
fully embedded device which transmits the temperature by radio signals. Any sensor with an access hole, stem or leads introduces some
degree of disturbance.

This problem illustrates the contrast between the alternative
attitudes.   One is to minimise the error qualitatively by using
fine-wire thermocouples, small access holes backfilled with
powdered refractory to cut down convection, laying the couples
along the expected isotherms and hoping that this will do.   The
other approach requires far more work; it consists of quantifying
the error.   This generally demands a subsidiary validating experiment.   Supposing we need to measure the temperature gradients in a
furnace lining.   We should start with the smallest possible access
holes and thermocouples at one representative spot, take temperature readings (remembering any resistance corrections), then en-
large the hole, using a larger, less delicate thermocouple. Finally,
a third attempt is advisable with an even larger set, to check that
the error is continuous.   By plotting the results against wire and
hole size we may be able to extrapolate back to an infinitesimally
small hole.   The errors may depend on depth as well as temperature,
therefore the experiment should be done at two or preferably three
different depths, successively.   If done at the same time, the
presence of a deep hole and long thermocouple may falsify the
readings in a shallower one close by.   This type of series gives
the additional advantage that the temperature at the inner surface
can be estimated by extrapolation, whereas a sensor breaking
through to the inside may have the wrong emissivity, etc., and give
a false result.

When measuring temperatures in good thermal insulators it is
important to ensure that the instruments do not consume or generate
appreciable amounts of energy at the sensor. Null methods and

intermittent testing may be necessary despite their slowness.

If a system is slow to respond and it is inconvenient or impossible to wait for equilibrium, it is possible to extrapolate. Temperature measurement may be used as an example. Suppose we insert a thermometer into a pocket formed in some object at temperature t_f. We then draw a graph of the instantaneous temperature reading t_i against time. The slope of the graph is found to be some function of the discrepancy or instantaneous error $t_f - t_i$. The author has found the slope to vary as $(t_f - t_i)^{1.1}$ to $(t_f - t_i)^{1.2}$. The reason for this appears to be that a thermal system of this kind involves at least one heat-path by natural convection, in the thermometer bulb itself or in the surrounding pocket of air. This heat-path conveys heat at a rate proportional to (temperature difference)$^{1.25}$. It is in series with a linear conductive heat-path and in parallel with another non-linear path by radiation. By taking several tangents, the law for the particular case can be found and hence the final temperature can be estimated.

2.10 SACRIFICIAL TEMPERATURE INDICATORS

The pottery industry has long relied on placing small cones or bars of selected composition in the furnace which sag or melt at predetermined temperatures. An extension of this principle is to identify the highest temperature reached by a surface during any process by putting on a range of witness marks of varying melting points. A proprietary range of crayons, metallic paints and stickers is available [5] by which the peak temperature can be bracketed within a few degrees by seeing which mark has fused into a smooth layer and which mark has retained its granular appearance, or at lower temperatures which sticker has carbonised and turned black. A range of paints with colour changes is also known but is more difficult to interpret since some of the colour changes are affected by time as well as temperature and by reducing or oxidising atmospheres. [6]

2.11 INDUSTRIAL INDICATORS AND RECORDERS

Many industrial pyrometers, indicating or recording, are based on moving-coil galvanometers. They are normally calibrated directly in temperature terms, with the appropriate thermocouple e.m.f. and resistance correction incorporated in the scale. If it is necessary to use an incorrect couple material or couple resistance, the actual value can be obtained by use of the principles set out in section 2.2.2 provided the meter resistance and the intended thermocouple values are stated (normally on the scale itself).

Some instruments use a delicate filament suspension which makes them sensitive to levelling. All are liable to damage from vibration and may need to be clamped and tied down for transport. If in doubt, the makers should be consulted for packing and commissioning instructions; a simple clamp intended for local transit may be insufficient for severe transport shocks.

Sensitive galvanometers are not always strong enough to drive an
ink pen. Low-friction recording mechanisms include a heat-pen
carrying a red-hot filament which burns a trace on a treated paper
chart without actual contact, an optical lever system by which an
ultraviolet light-spot burns a trace, or intermittent print mechanisms which leave the pointer free most of the time, clamping it at
perhaps half-minute intervals and printing a dot on a chart. Some
use a typewriter ribbon, others print through a carbon paper. The
latter version can give duplicate copies direct from the instrument
by using thin charts and extra carbon papers.

The majority of recording instruments use the Wheatstone bridge
principle, either to measure resistance or as a potentiometer to
measure e.m.f., using a standard reference voltage. A bridge
recorder needs an imbalance detector - a mechanism to detect the
direction of the imbalance and external power to move the slide-
wire. Electronic imbalance detectors have replaced the galvanometer type with its ingenious clamping and follow mechanism. The
imbalance current is amplified and drives a servo-motor which
balances the bridge and has ample power to spare for driving a pen,
digitisers, distant repeaters, controllers, etc.

Many recorders have a multi-channel facility, switching in up to
20 circuits in sequence. The switch is connected to a printing head
which types an identifying number against each dot on the trace.
Any number of points can be bridged together to the same input to
increase the printing frequency. Some recorders use multi-coloured
typewriter-style ribbons to identify separate channels. This is
useful when the traces run close together so that numbers would
overlap. Most multi-channel recorders do however provide a choice
of chart speeds which makes it possible to spread out the dots and
numbers.

The multi-channel recorder is used predominantly with thermocouples rather than resistance devices in view of the danger of
false resistances arising in the switches. If a temperature
differential is being recorded and the instrument as usual is cold-
junction compensated, a spare channel can be used to give a zero
trace. The instruction manual should be consulted on how to do this
- it often requires connection to a local thermocouple or shunting
out with a resistor rather than just leaving open-circuited. An
open circuit is liable to errors from leakage currents, false voltages as the switch opens, etc.

QUESTIONS

(Beware of trick questions) Use figure 2.6 or tables.

2.1 A copper/constantan thermocouple has a resistance of 10 Ω. It
is connected to a voltmeter of 80 Ω resistance. The couple is known
to be at 200 °C and the meter is at 20 °C. What will the meter
reading be?

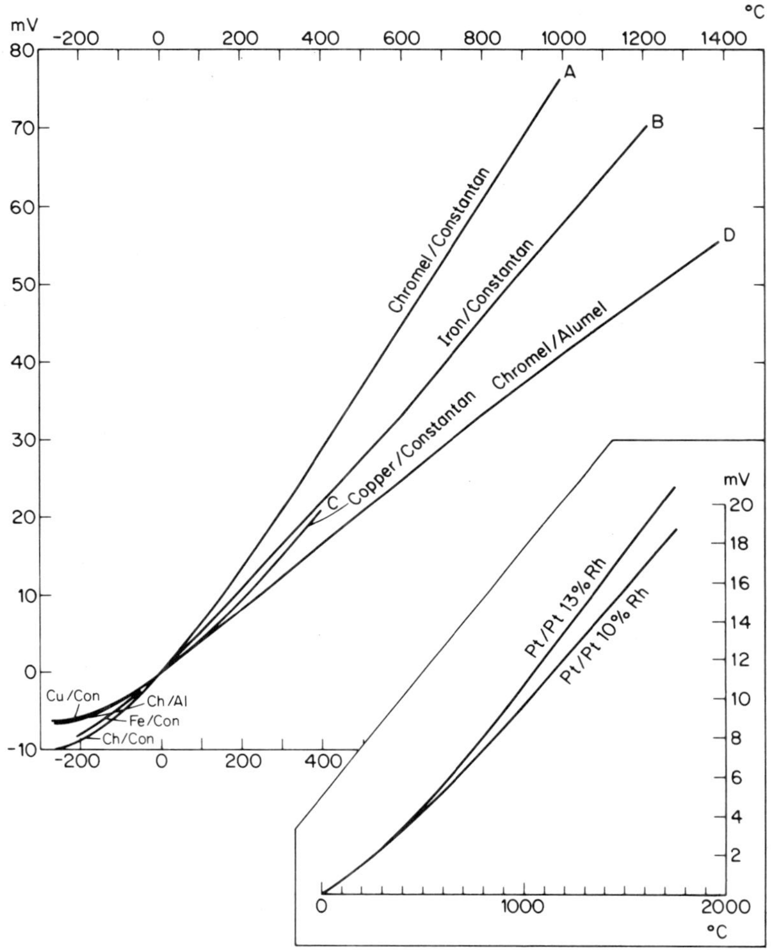

Figure 2.6

2.2 A potentiometric pyrometer is calibrated for Ni-Cr/Ni-Al
(Chromel/Alumel) and is cold-junction compensated by a zero-shift
mechanism. The pyrometer is at a temperature of 20 °C. Owing to a
shortage of supplies it has been connected to an iron/constantan
thermocouple and reads 325 °C. What is the actual temperature?

2.3 A pyrometer designed for couples of 5 Ω resistance is being
used with a couple of 10 Ω resistance. The meter resistance is
45 Ω and it is cold-junction compensated. What are the true couple
temperatures if the meter reads (a) 0 °C, (b) 20 °C, (c) 90 °C?
The meter is at 20 °C in each case. Assume a linear e.m.f./tempera-
ture relationship.

2.4   An instrument connected to a copper/constantan couple reads
1250 °C.   Is there anything wrong?

2.5   A potentiometer is connected to a Chromel/Alumel couple whose
hot junction is at 450 °C, connected as in figure 2.2b.   The junction box is at 60 °C.   What reading would you expect at the meter,
in millivolts, if the meter is at 20 °C and (a) the correct compensating leads have been used, (b) plain copper extension leads have
been used, (c) the correct compensating leads have been used but
connected the wrong way round?

2.6   Two iron/constantan couples are connected back-to-back to
measure a temperature difference, using a voltmeter of 120 Ω resistance.   The couples have a total resistance of 30 Ω.   What reading
in millivolts would you expect for a temperature difference of 20 °C
when the mean temperature is (a) 700 °C, (b) 200 °C, (c) -150 °C?
Would there be anything wrong in substituting a voltmeter of 20 Ω
resistance?

2.7   A mercury-in-glass thermometer intended to be immersed up to
the -20 °C mark is immersed correctly in a jar of ice and water and
is found to read 3 °C.   Suggest two possible reasons for this and
explain how you could check your hypotheses.

2.8   A total-immersion thermometer is known to be correct.   What
should it read if immersed up to the 10 °C mark in (a) ice at 0 °C,
(b) water at 80 °C, the exposed stem being at 15 °C?

2.9   A thermometer similar to that in question 2.7 is found to read
-0.5 °C when correctly immersed at a true 0 °C.   What is it likely
to read if suspended in a bath of condensing steam when the atmospheric pressure is (a) 760 mm of mercury, (b) 740 mm of mercury?
Remember to allow for zero error and for immersion error.

3. STATIC MECHANICAL PROPERTIES

It should be remembered that some standard tests and testing machines
are intended chiefly for production control, to ensure that the
material is as good as usual by taking periodic samples. In using
such test data for design purposes we must consider how closely they
relate to service conditions so that we can avoid wasteful overdesign
on the one hand, dangerous overstressing on the other.

3.1 TENSILE TESTS, SHAPE AND SIZE EFFECTS

A typical tensile testing machine is a reversed press, with a force-
measuring device on the upper cross-head or platen and a means for
pulling down the lower platen slowly. Many provide a facility for
making an automatic record of machine extension versus force, giving
a load-extension curve. This is of course the reverse of what
usually happens in practice, where the load is the primary cause and
the extension is the consequence.

   Tensile strength is measured on standard specimens of such length
that slip-planes can develop freely, generally over 4.5 times the
diameter or width, with enlarged ends to ensure that fracture occurs
away from the ends. In routine tests the extension is measured at
the machine cross-heads and is quoted as elongation of the active or
gauge length. For accurate purposes this is quite unsatisfactory,
since the machine extension includes not only the required extension
of the gauge length, G (figure 3.1) but also extension of the transi-

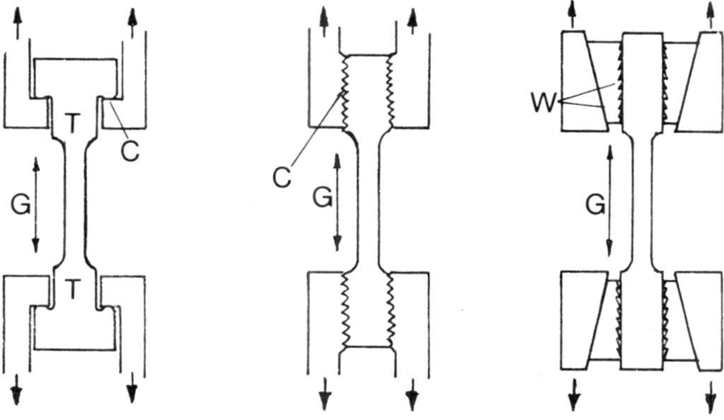

Figure 3·1

tion region, T, settlement at the contact faces, C, and quite
possibly some movement at the wedge grips, W. The true gauge-length
extension should be measured by one of the many suitable extenso-
meters in existence. In some machines the load may be subject to
friction errors. These can be estimated by a loading-offloading
cycle before the specimen has started to behave non-linearly. The
recorded load is not necessarily the load on the specimen but may
need to be corrected if it includes part of the grips which can be
quite massive compared with some small specimens. This is found by
taking a zero reading with the specimen holder *in situ*.

   The reasons for using a standard specimen shape are connected with
the behaviour of ductile materials. When permanent deformation starts,
plastic flow is uniform at first, increasing the length and diminishing the thickness. This process is only stable while the rate of
strain-hardening exceeds the rate at which the cross-sectional area
diminishes. Most materials eventually form a neck. The standard
specimen shape is long enough to allow this necking to take place
freely. A longer specimen would give almost the same results; the
chief differences would be a greater chance of a weaker region of
material coming within the gauge length. A shorter specimen would
inhibit the neck formation and in ductile materials will generally
show a higher ultimate stress value, tending towards the ultimate
true stress based on the final load and the fracture area. In
brittle materials or under cyclic loading, the short specimen will be
affected by stress concentrations in the transition between test section and end-fittings.

   There is an important lesson in this. Many engineering components
have short sections at high stress: in corners, at shoulders, at the
root of screw threads. If used under steady loads they get the benefit of this ductility effect, standing up to local 'theoretical'
stresses which would horrify the designer if he had investigated them
fully.
   In actual fact the local region yields and re-distributes the
load more evenly. If an existing design is converted to using a
more brittle material, this self-relieving feature may disappear.
A mere substitution of a stronger material does not always produce
a stronger component.

   If specimens of standard shape but reduced size are used, this
alters the conditions of plastic flow, since the number of grain
boundaries in a given plane is reduced. It also reduces the chances
of securing a typical sample and increases the effect of load-
measurement errors. On the other hand, taking a large number of
small samples produces information on the consistency of properties
from place to place, which is averaged out in a large sample.

   Tensile testing does not detect longitudinal slag inclusions in
rolled material unless the specimen is cut in the appropriate direction *and* happens to be taken from the right place.

3.2 COMPRESSION TESTS

These are not widely used in metals; in short specimens they are

complicated by friction at the loading faces, in long specimens by
lateral deflections. Their main use is in assessing the crushing
strength of timber and concrete. Unlike tensile testpieces, compression specimens are anything but self-aligning; the initial
alignment is very important and any loss of alignment during the
test should be noted and referred to in reporting the results. When
carried out in simple hydraulic presses, friction errors may be
significant. If the test is done in a tensile machine, the weight
of the additional fittings may need to be allowed for.

3.3 SHEAR TESTS

These simulate shearing or cropping action and are relevant to designing riveted joints, shear-pins, keys in shafts, etc. The shear
strength in this restricted situation is usually higher than the
yield stress in torsional shear.

We may reflect on the problems of engineering language when the
use of the terms shear and shear stress has become so remote from
the action of a pair of shears that the latter needs to be re-
defined. The A.S.M.E. pressure vessel code uses the term 'restricted
shear'.

The test is best carried out with the specimen shearing on two
planes, for symmetry (figure 3.2).

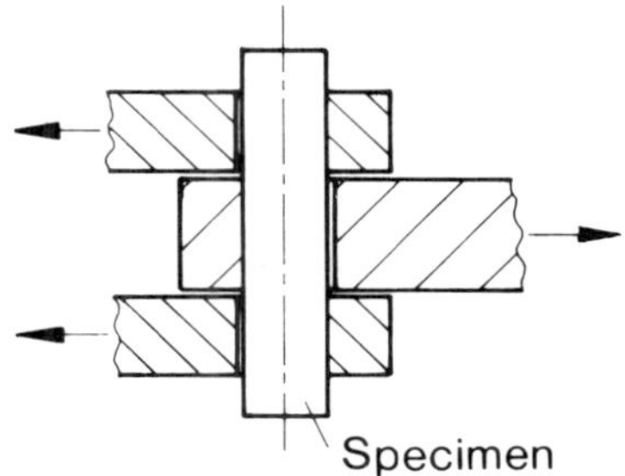

Figure 3·2

3.4 BENDING TESTS

These are obviously of some relevance since bending stress conditions
occur in most engineering components, often providing the most
serious stresses. In many cases bending behaviour can be deduced
from tensile data, particularly when loads are steady. There are,
however, many materials with layered or unidirectional properties,
for example, sandwich sheets such as Alclad, stainless-clad carbon
steel, temper-rolled sheet, surface-hardened gears, shot-peened
springs, materials weakened by grinding heat or decarburisation,
flake cast iron in which the graphite-filled cavities extend readily
but resist compression, and particularly fibre-reinforced materials
in which the tensile face develops the full strength of the fibres,

while on the compression side the fibres are imperfectly supported
by the low-modulus binder and can buckle at low stresses (figure 3.3).

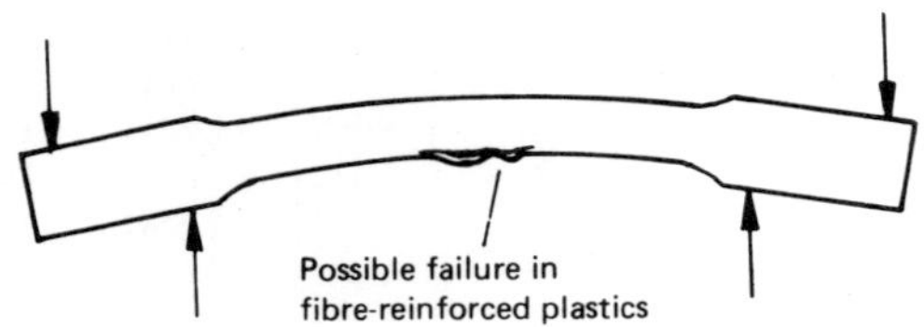

Figure 3·3

It needs little imagination to see that bending tests are liable
to suffer from friction errors and from uncertainties about the true
bending moments. Also, students are liable to overlook the distinc-
tion between bending of narrow bars and of wide plates. In the
latter case the elastic deflections are more truly represented not
by E but by $E/(1 - \nu^2)$, often written as E'.

A typical three-point bending set-up is shown in figure 3.4.
Only the most unthinking will confuse the rounded loading points
with *frictionless* rollers; the rounding is to prevent surface
damage. Significant friction errors are to be expected. At small
deflections the convex surface gets longer while the supports stay
the same distance apart; thus the contact zone tries to prevent the
surface from extending. This problem is rare because the loading
device is generally somewhat flexible.

At larger deflections the two loading-points slide relatively
outwards, so obviously the friction component now acts outwards,
adding to the bending moment. The equations are shown in figure 3.4
for each case. To estimate the friction coefficient, μ, we may use
the loading-unloading routine. If the force at a given deflection
is P_i during the increasing phase and P_d during the decreasing phase,
$\mu \tan \theta = (P_i - P_d)/(P_i + P_d)$. For uniform beams, $\tan \theta \simeq 3\delta/L$.
The notation is as in figure 3.4.

In four-point bending, using two inner supports instead of one,
the situation is fundamentally similar but even more confusing, since
friction and radius effects occur at all four points.

In view of the problems associated with beam tests as above, it
is tempting to resort to cantilever tests (figure 3.5). These
undoubtedly reduce the uncertainties of bending moment but for finding the elastic properties they require correction. However firmly
we clamp the end of the specimen, some deflection takes place within
the clamped part so that the cantilever has an additional effective
length. Since end deflection for a given load varies as the cube of
the length, this end error can be quite important. An experimental
approach to finding the length error is to repeat the experiment
with several different lengths.

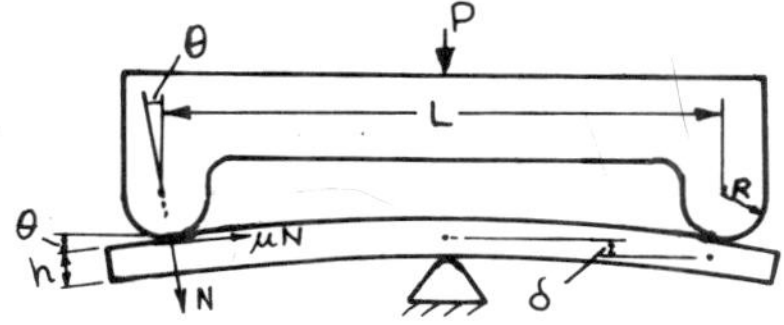
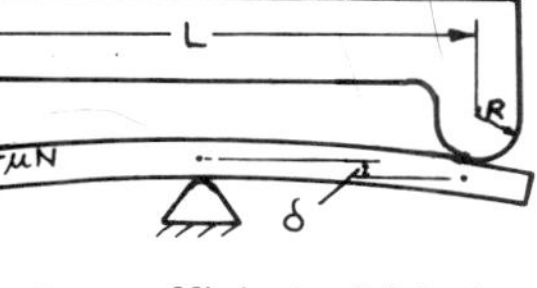

Load increasing:

$$P = 2N \cos \theta - 2\mu N \sin \theta$$

therefore

$$N = \tfrac{1}{2}P \sec \theta/(1 - \mu \tan \theta)$$

$$M = \frac{P}{2} \cdot \frac{\dfrac{L}{2} + (\delta - R - \dfrac{h}{2})\tan \theta - \mu(\dfrac{\delta}{2} + \dfrac{h}{2})}{1 - \mu \tan \theta}$$

For uniform beams, $\tan \theta \simeq \dfrac{3\delta}{L}$

δ = deflection

μ = coefficient of friction

M = bending moment at centre

(a)

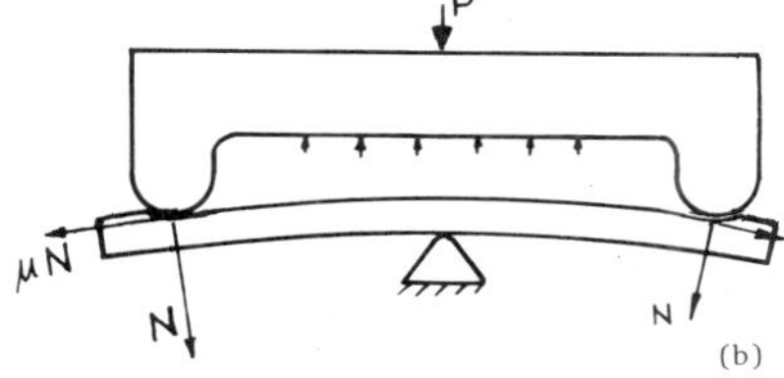

Load decreasing:

$$N \simeq \tfrac{1}{2}P \sec \theta/(1 + \mu \tan \theta)$$

$$M = \frac{P}{2} \cdot \frac{\dfrac{L}{2} + (\delta - R - \dfrac{h}{2})\tan \theta \div \mu(\dfrac{\delta}{2} + \dfrac{h}{2})}{1 + \mu \tan \theta}$$

(b)

Small deflections ($\delta < \tfrac{1}{2}h$)

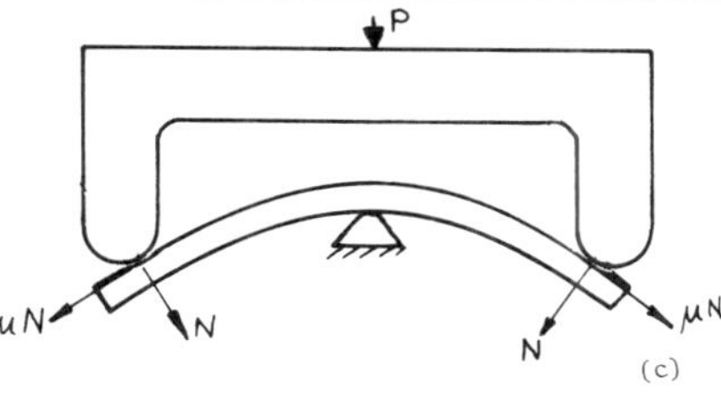

Load increasing:

Equations (b) since
friction acts outwards

(c)

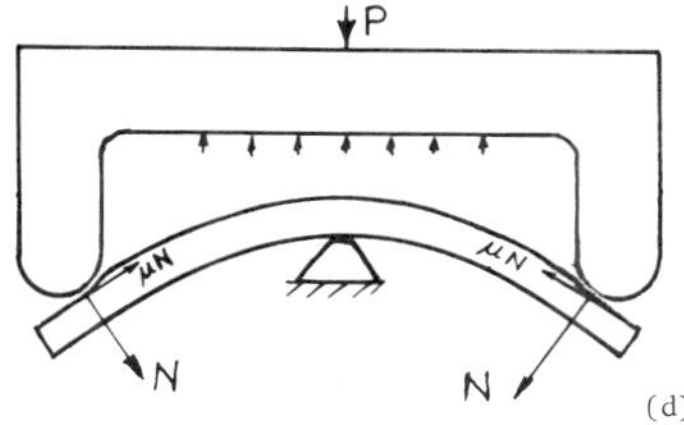

Load decreasing:

Equations (a) since
friction acts inwards

(d)

Large deflections ($\delta > \tfrac{1}{2}h$)

Figure 3·4

Supposing we wish to find Young's modulus, let us take a long strip of constant width b and thickness t and clamp it so that an apparent length L_a protrudes. The additional active length within the clamp, L_c, presumably depends only on the thickness. At the load end there is a further (small) error presumably proportional to end slope. The deflection at the load point $\delta = W(L_a + L_c + k\theta)^3/(3E'I) + WL_a/Gbt$, (bending and shear). W, L_a and I are known, δ and θ are observed each time, leaving three unknowns E', L_c and k. If

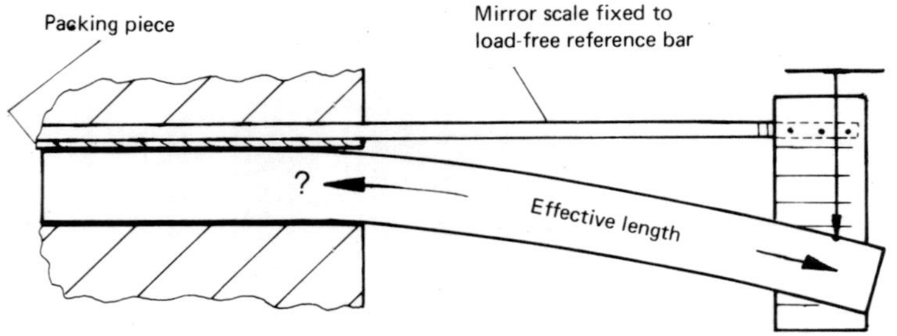

Figure 3·5

the shear is negligible, three experiments should suffice to eliminate the unknowns, with a fourth as cross-check. The loads would be
chosen such that the apparent bending moment at the fixing is constant, in other words $W \propto 1/L_a$.

It may be of interest here to note that material in sheet form
is not necessarily of uniform thickness as supplied. Each specimen
should be measured in several places. Sheet plastics are liable to
greater variation than sheet metal.

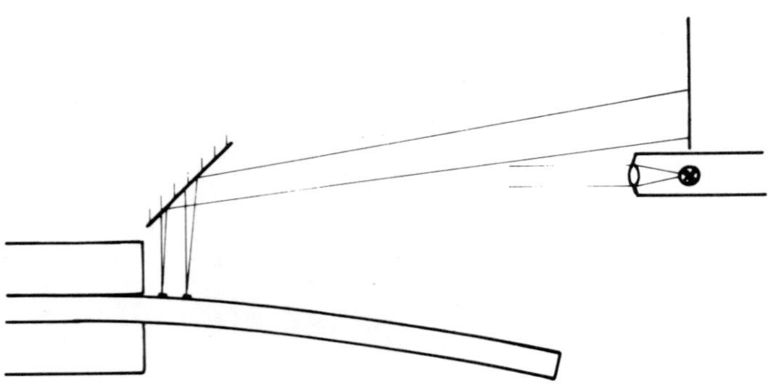

Figure 3·6

If there is the slightest doubt about rigidity of the mounting,
it is a good idea to place a reference bar in the same clamp.

A more refined approach is to stick small pieces of mirror to the
member and find the deflections by a light-beam. The beam changes
through *twice* the angle change at the mirror (figure 3.6).

Plastics, especially thermoplastics, have several peculiarities
to be considered in testing: non-linearity, creep, memory, temperature dependence and in many cases strong dependence on absorbed
moisture. Non-linearity is taken care of by defining Young's modulus
at any particular stress in two ways, either the tangent (incremental)
modulus or the secant modulus as from zero load, shown by the lines T
and S in figure 3.7. Creep makes it necessary to establish a consistent testing procedure, allowing enough time at any given setting for
internal heating to dissipate itself, initial settlement to cease and
either keep long-term creep errors negligible or apply a correction
by noting the time-deflection curve. Ambient temperature should be
observed for future reference.

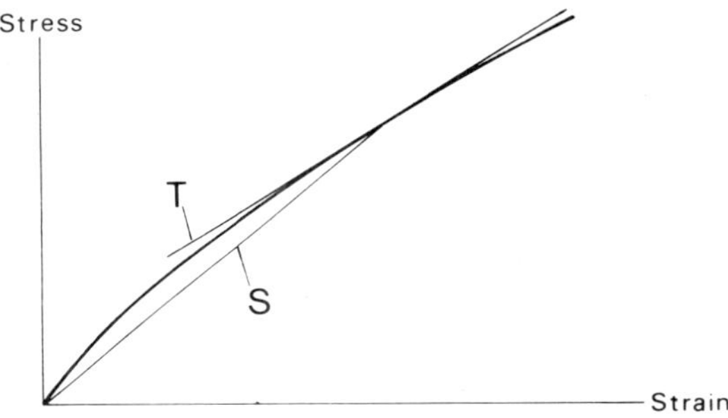

Figure   3·7

When testing thermoplastics we may get peculiar effects due to
memory - a form of delayed creep. If straight material is kept
flexed for a short time, it will recover almost fully. If, however,
it is bent in some shape for a long time, then straightened and later
bent to another shape, on release it will tend to go back through its
history towards the earliest long-term shape. This is very notice-
able when plastic tube or strip has been kept in store in the form of
coils.

Most materials become less rigid as they get warmer, E and G decreasing with rising temperature. A notable exception to this is
natural rubber. At extreme conditions of cold, rubber goes very
rigid, when very hot it melts. In between, however, say between 0
and 70 °C, its moduli are proportional to the absolute temperature,
so that the force exerted at any given deflection goes up with temperature. This applies to tension, compression or shear and is quite
distinct from the effect of thermal expansion. The effect can be removed or reversed by adding large amounts of reinforcing filler to
the rubber mix. There may be some applications for this thermo-
elastic effect in instrument design, as a compensator for other thermal effects.

QUESTIONS
Assume elastic deflections.

3.1 A tensile testpiece as in figure 3.1a has a gauge length of
50 mm, of 8 mm diameter; then it changes smoothly over a length of
5 mm at each end into enlarged portions of 16 mm diameter, each 20 mm
long before the heads are formed. Estimate the ratio of total elastic
extension to gauge length extension. Ignore effects at the contacts C.

3.2 A material of Young's modulus E and Poisson's ratio ν should
have a modulus of rigidity G given by $E/2(1 + \nu)$. Assuming a cantilever of square section, thickness t, length 3t, calculate the ratio
of shear deflection to bending deflection if ν = (a) 0.29, (b) 0.4.
For your convenience, the usual formulae are given below, for a section of width b, thickness t.

$$\text{Bending deflection} = \frac{\text{load} \times \text{length}^3 \times 12}{3Ebt^3}$$

$$\text{Shear deflection} = \frac{\text{load} \times \text{length}}{Gbt}$$

3.3 A *wide* cantilever specimen, 2.5 mm thick, 20 mm wide, 50 mm
exposed length, is made of a brass of $E = 10^5$ N/mm^2, $\nu = 0.35$. Find
the deflection under a load of 250 N (a) using the apparent length,
(b) assuming an effective length of 51 mm.

3.4 A uniform beam, 20 mm wide, 1 mm thick, is loaded as in figure
3.4, where L = 120 mm, R = 3 mm, μ = 0.25. The material's E' = 10
N/mm^2. Find approximately the load required to produce an increasing deflection of 6 mm and also the load that will just support this
deflection during unloading. Use the usual beam formula to establish
the bending moment.

4. DYNAMIC MECHANICAL PROPERTIES

This chapter discusses fatigue testing, fracture testing and general rapid loading.

4.1 FATIGUE TESTING

This subject has numerous problem areas, some well-established, others sometimes ignored.

4.1.1 *Stress-gradient or Size Effect*

In cases where the load is tensile only, uniform throughout the region under test and fluctuating only in magnitude, a small tensile-type specimen survives substantially the same stresses as a similar large specimen. (If, however, the load is partly compressive, then any given misalignment in the machine has a greater bending effect on small specimens.)

This independence of size is lost in bending specimens or specimens with stress-raisers such as shoulders, screw threads, keyways or holes. In such a specimen the highest stresses are localised. Under cyclic loading the cracks propagate in small steps and will stop at the earliest grain boundary at which the stress is low enough. Accordingly the depth to which a crack penetrates depends on whether the stress level falls rapidly in terms of grain-diameters.

It is well established that small specimens with non-uniform stresses show unrealistically high strength, expressed usually as a *low* effective stress concentration factor, which is dangerously misleading when used for design purposes. In large machinery a photoelastically determined or theoretical stress concentration factor is a truer guide than a small-scale experiment. This problem should be kept in mind in failure investigations. Specimens cut from failed components are of necessity smaller than the originals and are likely to give misleading results (see below for a fuller discussion). This also casts doubt on some of the older stress concentration data published before the size effect was widely understood; the specimen size and grain size were not always reported.

4.1.2 *Prior Load Effect*

A ductile component with stress-raisers is likely to yield locally at the first loading with the local *stress* staying fairly constant while the local *strain* increases, with slight change of shape. When the load is removed the component is only slightly distorted but the local yielded material is under a residual stress, being forced back towards its original size by the elasticity of the surrounding

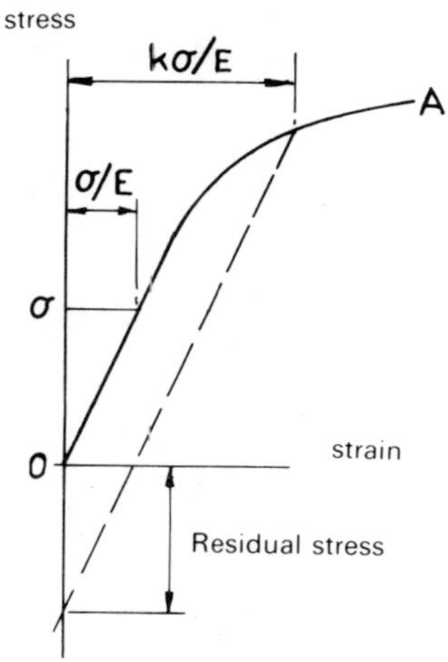

Figure 4·1

material.

Consider figure 4.1. OA is the stress-strain curve for the
material. A load is applied to give a general stress σ, general
strain σ/E. Locally the strain is magnified by some factor k to
kσ/E. The factor k is presumably somewhat similar to the elastic
stress-concentration factor. When the load is removed, the yielded
region is under a considerable residual stress which affects subsequent behaviour. If subsequent loads are in the same direction as
the first load but lower in magnitude, the peak stresses are
greatly reduced. Subsequent reverse loads on the other hand give
augmented stresses.

This is important in structures which are given an overload test
(proof load). Most parts of such structures are greatly strengthened
by the test, but those parts which later receive reverse stresses
are weakened. Reversing stresses occur due to wind loads on masts
and cranes, rolling loads on bridges, temperature changes in pressure
vessels, etc. For further discussion see appendix 1.

What is all this to do with fatigue testing? Consider the dili-
gent student investigating a component - say, a welded joint. In
setting up a fatigue test, the component is clamped in the machine.
To make sure it is well secured, what is more natural than to give
it a good pull before starting the tests?

Again, consider a pressure vessel to be fatigue tested. It is
filled with water, all openings blanked off and connected to the
pressure source. Before the engineers arrive to observe the tests,
the fitters want to make sure the connections do not leak, so they
give it an unreported prior test.

A result of this effect was that the Comet Mk.I fuselage design

was apparently strong enough in fatigue. A specimen structure was
successfully subjected to a large number of pressure cycles repre-
senting cabin pressurising in high altitude flight. Unfortunately
the same specimen had previously received a large static proof test.
Fuselages put into service did not have the large static proof test,
only a much less severe one. In consequence they had much less of
the favourable residual stresses and broke up in service due to
local fatigue. [7]

Similarly, any fatigue result may be confused and invalidated if
it was preceded by a high static loading not part of the programme.
An example is given by Taylor, [8] see appendix 1.

4.1.3 Machine Effects

There are four basic layouts of fatigue tests: rotating beams,
oscillating beams, push-pull and torsion. The rotating beam method
is the most straightforward, the chief troubles being the obvious
ones of (a) the specimen becoming bent as a crack develops and (b)
faulty bearings giving jerky motion.

It is interesting to compare the effects of a developing crack
in a testpiece and in a working component.

In push-pull testing a crack will produce increasing *stress* if
the force applied is the same at each loading. If, however, the
deflection applied is the same, the crack will permit the deflection
to develop with *less* force being needed; the stress may get lower.

In rotating tests, the inertia may be great enough to keep the
specimen straight so that a crack may reach the neutral axis and
then stop growing, giving an artificially long survival.

In real life, it all depends on the degree of support. In many
cases a crack will simply mean a reduced cross-section still carrying the full load. If, however, the structure is redundant, with
multiple supports, a crack may well result in transferring the load
away from the cracked section as a result of the increased deflection of the cracking member. It is often considered wise to design
a structure in a sub-divided way with redundancy of support. There
are many examples of this, the most basic is wire rope as opposed
to chain (members in parallel rather than in series).

In push-pull and torsion testing it is tempting to neglect dynamic effects. It is usual to measure the force by connecting the
specimen to a dynamometer (load cell, weigh-bar); the force is
applied via this to a fixed specimen (figure 4.2) or via the specimen to the dynamometer (figure 4.3). The force recorded is *not* the
force on the specimen, but the force on the specimen plus or minus
the force to accelerate the specimen holder. With small specimens
and massive holders the error is appreciable, especially at high
speeds. In case 4.2, the correction will depend on speed and specimen flexibility, in case 4.3 on speed and dynamometer flexibility.
In torsion work, the flywheel effect of the specimen holder may be
quite large compared with the polar moment of a small specimen.

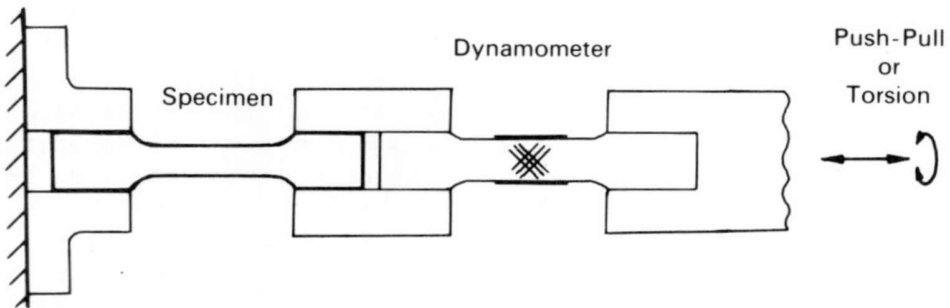

Figure 4·2

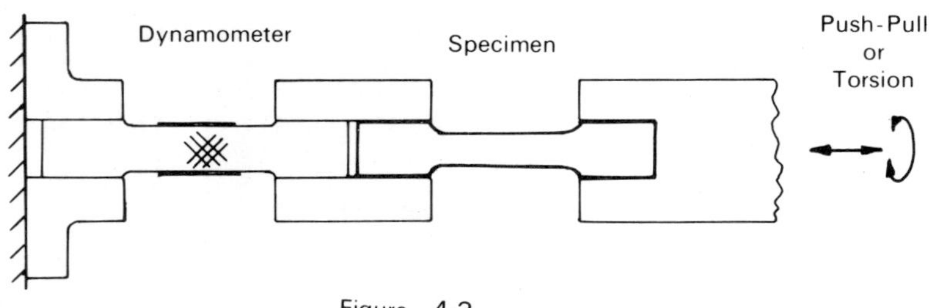

Figure 4·3

The type 4.2 set-up is sometimes calibrated by a dummy specimen
fitted with strain-gauges. Only if the dummy is of the same stiffness as the working specimen in the same modes will the calibration
be valid. Type 4.3 is easier to calibrate; the percentage error
depends only on speed.

If a strain-gauge is not fully stuck but has air bubbles in the
glue film, this may have totally different effects in tension and
compression, or in shear which may occur in combined tension-
torsion tests. One way of detecting bubbles is to protect the
gauge with thin flexible plastic and apply local pressure with a
small rubber while the gauge is connected. The flexure strains
due to squashing the bubbles, should give large output signals.
The author is indebted to D. A. W. Taylor of Sheffield University
for this item. See also section 6.8.

The importance of such errors is this: if the real specimen
loads are less than the apparent machine values, the specimen will
appear to stand higher stresses than it really can; if used as
design data this will be dangerous. In the reverse arrangement the
specimen will be loaded more than it seems, the results will tend
towards safety in design. However, in a failure investigation the
opposite is true; a specimen from a broken part, tested, say, by
method 4.3 without correction, will fail prematurely and cause the
blame to be put wrongly on the material, perhaps concealing an
important design lesson.

In testing for limited fatigue life at repeated high stresses, the property of the material under study is crack propagation. The question whether a crack will grow at a given level of stress and how fast it will grow is concerned with the release of strain energy in the structure and its transfer into the crack zone. A typical test arrangement consists of a plate with a pre-formed notch or

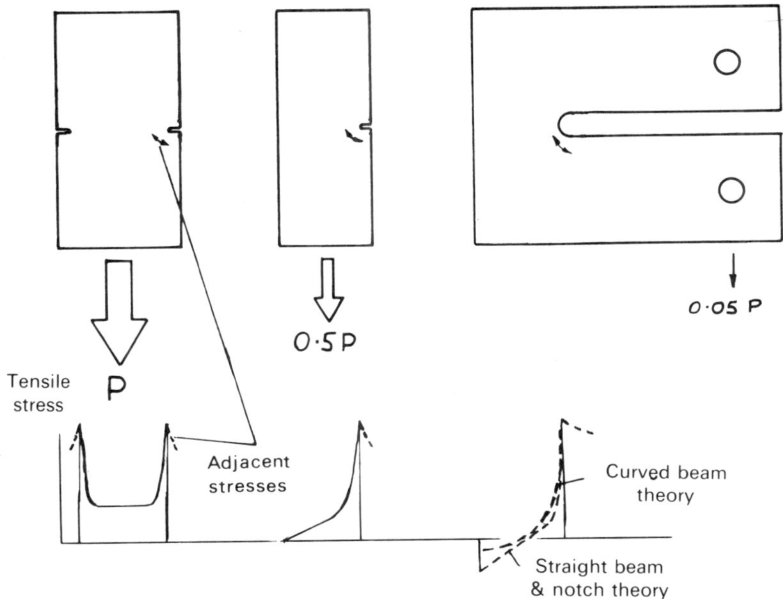

(a)   Approximate loads for equal maximum stresses on equal areas

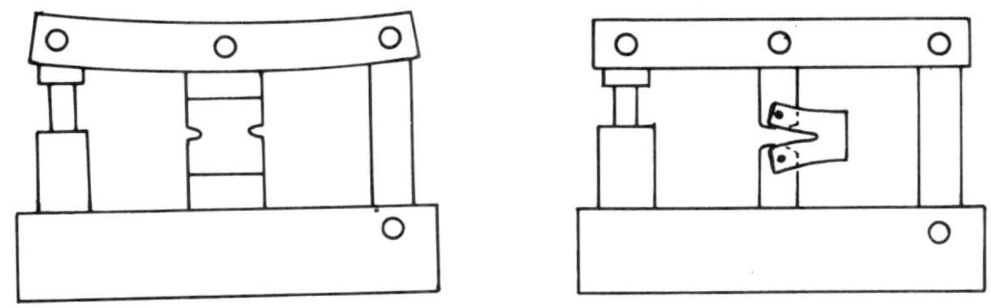

(b)

Figure   4·4

crack, subjected to cyclic loads, as in figure 4.4. It has been
noted that the behaviour depends on plate thickness, thicker plates
failing faster at a given level of stress.

If the testing apparatus were completely rigid, the energy trans-
fers would be confined to the plate specimen only; in that case it
might be difficult to explain the dependence on the thickness, since
the strain energy release and crack-growth energy per step should
both be proportional to thickness, the only variation being due to
grain size in the metal.

When we consider the effect of the testing machine also, it is
possible to argue that energy release from a given machine depends
on force change *and* deflection, each being proportional to thickness. Thus the energy content of the same machine when testing
specimens of various thicknesses will be proportional to the *square*
of the thickness, while the specimen energy only varies linearly
with thickness. The thicker the specimen, the greater is the
machine effect. It should be noted that fracture problems at
unexpectedly low nominal stresses have been noticed in large uniform structures such as ship's hulls where the stress levels would
be relatively uniform over a large region, making large amounts of
strain energy available for release. This would correspond to a
test set-up where the specimen thickness is comparable with the
thickness of the load-bearing parts of the testing machine. Such a
set-up would seem unscientific inasmuch as it does not isolate the
material property from the structure property but it would be much
more realistic (see appendix 2).

4.3 RAPID LOADING

Measuring strength at very high loading rates is carried out by
impact tests and sometimes by other rapid loading methods. The
usual engineering impact test consists of hitting a standard specimen with a pendulum and observing the energy absorption. This type
of test is purely comparative.

Impact tests are open to one major criticism. The speed of
loading (strain rate) is higher than would generally occur in
machinery, thus conditions are almost adiabatic. Though the elastic energies are small, during yielding prior to fracture the local
temperature rises could be quite significant, when perhaps 50 joules
are concentrated into the central part of a specimen. This could
give a misleading idea about the temperature at which brittle frac-
tures are to be expected in service, since in service the local
temperature could be appreciably lower than in a very fast test.
Such discrepancies have in fact been noted.

The methods used to give quantitative values of high strain-rate
strength and elongation fall into two types

(a) hammerblow tests monitored by strain-gauged dynamometers such
    as in Hopkinson's bar;

(b) test rigs based on gas explosions.

The hammerblow type is usually arranged with the dynamometer
directly adjacent to the specimen. This should avoid serious errors
provided that the build-up of stress-waves can complete itself
before fracture and provided the connector is of low mass.

Gas pressure tests have appeared in research literature; the
problem here is more likely to lie in the interpretation than in
the experiment. A gas explosion wave is of low mass compared with
the containing structure, while the pressure rise is likely to be
spread over many microseconds or indeed milliseconds. The natural
period of a responding system such as a piston and piston rod, or
a pipe under internal pressure, is readily calculable and is often
of the same order, so that the loading is progressive. Some
researchers assume that a rapid load automatically doubles the
stress, and would have us believe that the material used has twice
the usual strength when loaded rapidly.

The points to be borne in mind are that a classic sudden load
free from impact is the only case giving 100 per cent overshoot
(peak forces twice the steady forces). Such a case is rare in
practice. Most cases either involve an approach velocity giving
the loading a hammerblow character with considerable mass effect,
or they are semi-progressive with only slight overshoot. Sometimes
sudden force changes affect only part of a system, so that 100 per
cent overshoot applies only to the sudden part of the load change
and only to part of the whole system.

QUESTIONS

4.1 A torsional fatigue test uses a specimen of 10 mm diameter,
effective length 35 mm and a dynamometer bar of the same material
of 16 mm diameter, effective length 60 mm. The connector, together
with the specimen end-pieces, acts like a cylindrical body of 80 mm
diameter, 1 kg mass. $G = 70000$ N/mm^2. Estimate the torque in the
specimen when the dynamometer torque is ±4 N m at 4000 cycles/
minute with layouts as in figures 4.2 and 4.3. Resonance may be
neglected. Be very careful about newton metres, millimetres, etc.,
when finding inertia torques.

4.2 A test as in figure 4.2 uses a connector of 0.25 kg mass to
test specimens of titanium of 1 mm diameter in pulsating tension at
tensile stresses cycling between 200 and 800 N/mm^2. If the test
frequency is 6000 cycles/minute, find the maximum force needed to
overcome inertia of the connector, for specimens of 20 mm and 100
mm length. $E = 11600$ N/mm^2 for titanium. What stresses in the
specimen correspond to these forces? Also estimate the resonant
frequencies.

4.3 A test rig consists of two steel beams ($E = 200\ 000$ N/mm^2),
600 mm long, 120 mm deep and 40 mm wide (see figure 4.5), connected
by rigid pillars P and a hydraulic cylinder C. The rig is used to
load specimens of 25 mm width to a mean stress of 200 N/mm^2. There
are three types of specimen: 1 mm, 2 mm and 10 mm thick. Making
suitable assumptions

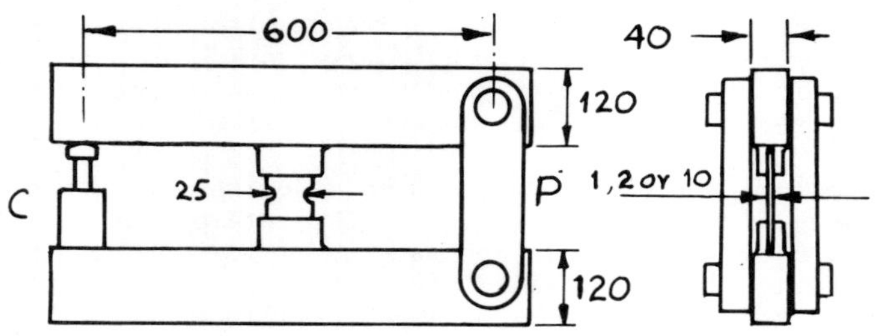

Figure 4.5

(a) Find the deflection of the centre of each beam relative to its ends and the energy stored in the test rig.

(b) Find the energy changes in the beams if the above specimens extend by 0.05 mm over all while the hydraulic cylinder remains position-fixed.

5. FLOW MEASUREMENT

5.1 QUANTITY AND TIME METHOD

The quantity and time method of flow measurement is absolute in the
sense of being free from assumptions about the relationship between
an observed pressure drop and the mean velocity or effective flow
area to which it refers. It does, however, conceal fluctuations in
flow rate; sometimes it is wise to combine it with a means for indicating velocity. The student is likely to meet a system shown
schematically in figure 5.1a. The flow circuit has an open section
from which a measured quantity is collected over a known period.
If this period is timed by stop-watch, it is useful to test the
system (operator + stop-watch) against a known master clock over
long and short periods, to train the operator, prove the time-
keeping of the stop-watch and establish the repeatability, that is,
the amount of random variation to be expected.

The beginning and end of collection involve short periods of
partial collection, as depicted by the graph. We tend to form
systematic habits in the timing of such actions and it is useful
to bracket the degree of uncertainty by measuring the minimum time
t_2 and the maximum time $t_1 + t_2 + t_3$, using two observers.

Other small errors come from liquid sticking to the walls, thermal expansion, air bubbles, etc., if volume is measured. Many of
these errors are eliminated by weighing the collecting vessel before
and after each fill. For repetitive work standard tinplate cans or
plastic drinking cups can be used.

If it is feasible to use two vessels, each mounted on a weighing
machine, with a rocking funnel, as shown in figure 5.1b, feeding
each vessel in turn, then the end errors due to incomplete collection disappear; over a large number of cycles the timing errors
are eliminated except for the first and last in the whole series.

Often it is sufficient to use tanks of known cross-sectional
area with a float or sight-glass to read the level. If flat-sided
tanks are used, a small error can result from bulging of the walls
due to fluid pressure, also from reduced cross-section due to
sludge, algae or corrosion products. The latter effects are not
constant; they increase with time, or may decrease due to maintenance. The tank method also suffers from temperature error and
evaporation error.

Gas quantities may be measured in constant-pressure gasometers
using floating bells with a compensator for the immersed part of the

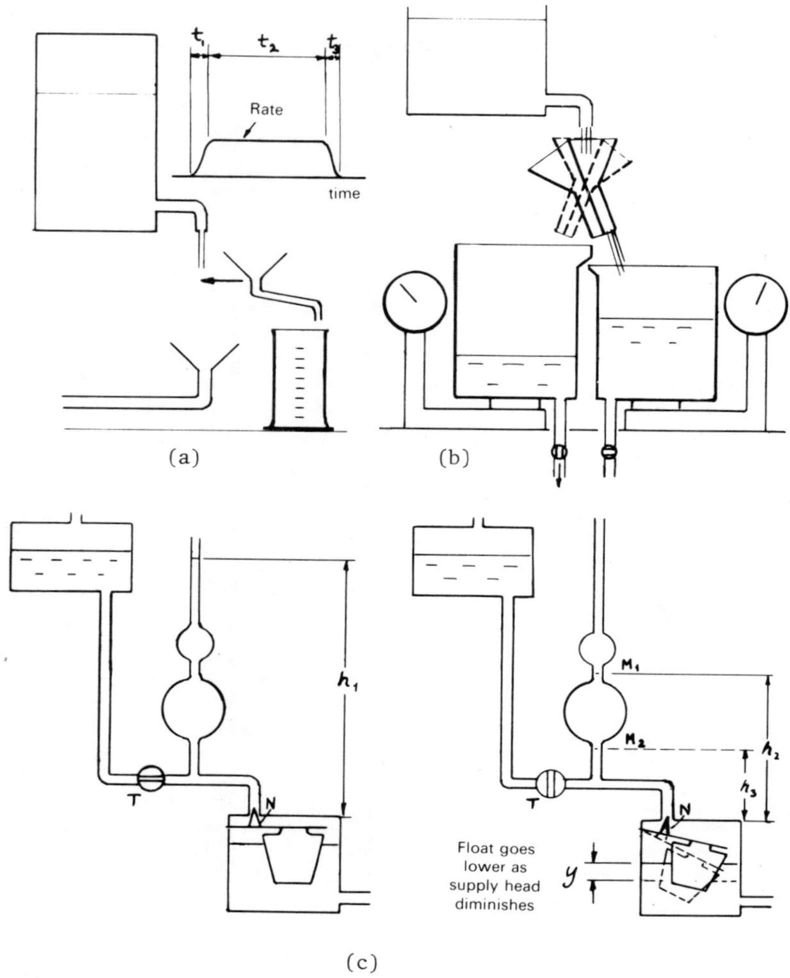

Figure 5·1

shell, or in rigid containers by carefully measuring the pressure
change, not forgetting the temperature also.

A variable storage error may occur on many test-beds but is not
generally known. In engine testing, the fuel flow rate is often
measured batch-wise by timing the passage of 50 or 100 ml. The layout is shown schematically in figure 5.1c. During normal running
the fuel supply head is h_1; when measuring, tap T is closed and the
time for the fuel to pass from mark M_1 to mark M_2 is noted. During
that time the fuel supply head falls from h_2 to h_3. The carburetter
float controls the fuel flow to suit engine demand by partly closing

40

the needle valve N. As the supply head falls, the needle valve has
to open further; the fuel and float level get lower during this
time, by an amount y. Thus the engine has consumed not only the
measured 50 or 100 ml but also an amount A × y, where A is the surface area of the whole float chamber, perhaps 20 cm^2 or more. The
amount y is not known. On the occasion which brought this to the
author's notice, the fuel line was partly obstructed; the normal
supply head h_1 was sufficient to maintain normal running but the reduced head between h_2 and h_3 was not, so that y was about 3 cm; the
actual consumption in a given time was not 100 ml but 160 ml, giving
apparent thermal efficiencies of some 45 per cent.

A rotameter-type meter in the fuel line working under steady-
state conditions avoids this error.

It should be noted that this is a typical systematic error,
always behaving consistently. A small error of this type could
easily remain undetected and lead to false conclusions; only when
the error is large does it arouse suspicion.

There are several meters based on the quantity displaced. Most
instrument manuals describe their construction. They are generally
supplied with a calibration. In most uses they can be considered
accurate, reading *very slightly* low due to through-leakage. Wet-
type gas meters are very sensitive to correct filling level and
correct inlet and outlet pressure but are free from through-leaks.

5.2 FLOW MEASUREMENT BY CONSTRICTION

This is the commonest method, less accurate than total collection
but much more convenient, and very suitable for remote reading and
flow control.

If the flow to be measured is obliged to pass through a constriction, its velocity must increase; the consequent pressure drop is
used for measurement.

At very low Reynolds numbers, the constriction can be a small-
bore tube. The pressure drop is the viscous loss plus the kinetic
head loss at the exit. If the pressure drop is high, correction
may be needed for temperature rise and reduced viscosity. [9] If
the fluid is non-Newtonian such as slurry, dust-laden air, water
containing long chain polymers, blood, etc., it may exhibit the Toms
effect and have abnormally low friction. Such fluids may also have
peculiar effects on the pressure tappings; one observer noted *high*
readings apparently due to impact effects, up to double the true
value. [10]

At higher Reynolds numbers, the usual restriction forms are
flared intake from free space, streamlined nozzle, venturi and
orifice. Empirical discharge coefficients and correction factors
are available for all these, in some standard textbooks or in
references 11 and 12. Figure 5.2 shows the commonest designs and
typical discharge coefficients.

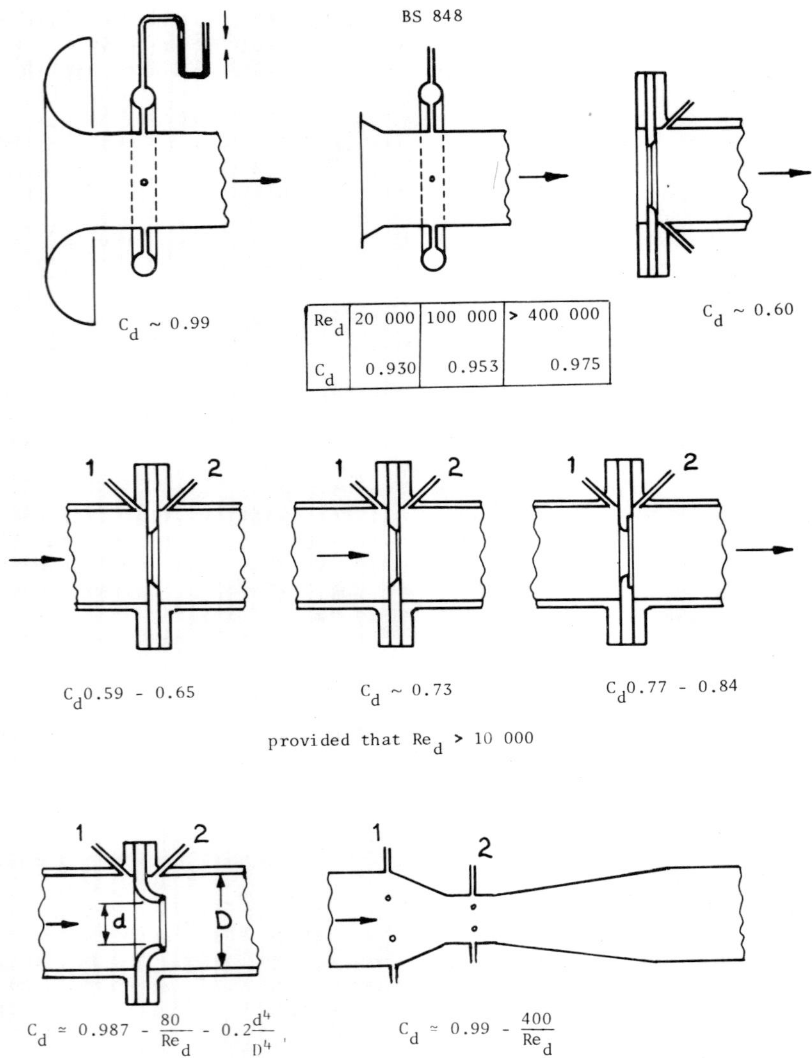

Figure 5·2

The pressure drop (signal pressure) is measured between stations
1 and 2. For orifice plates, two alternative tapping layouts are
available. The corners where the flow is relatively stagnant give
the best results but the other tappings are easier to make and are
often found in practical installations. In intakes from free space,
tapping 1 is the general surrounding static pressure (air or water,
etc.). It is often advisable to collect the pressure all round as

42

a precaution against local flows giving a false reading. In a good
pipe installation with well-straightened approach flow this ought
to be barely necessary but with intakes from free air or water it
is very important.

Venturis give lower over-all line pressure losses than orifices
for a given signal between 1 and 2, possibly as low as 20 per cent
of it (80 per cent recovery). Orifices recover ~ 30 to 60 per cent.

The choice is influenced by cost, simplicity of installation,
required signal (pressure difference) and acceptable over-all pressure loss.

The orifice with the sharp edge upstream is the commonest and is
believed to be the most predictable. Details of construction and
calculations would occupy a book almost as long as this. [11, 12]

The practical errors arising in use are

(a) back-to-front installation
(b) build-up of sludge or dust, altering the size and profile; this
     speeds up the approach flow or improves the streamlining, giving
     reduced readings for a given flow rate
(c) swirling flow which will give a centrifugal pressure distorting
     the wall static pressure
(d) leaky connections, especially with plastics tubing which
     gradually goes slack, or with high pressure lines
(e) moisture condensed in connections (for measuring flow of steam
     special condensing arrangements are necessary)
(f) air trapped in liquid-filled pipes (use transparent pipes if
     possible)
(g) temperature differences in tapping lines
(h) buoyancy effect of fluid itself
(i) pulsating flow
(j) insufficient settling length before the constriction.

Errors of type (d) can be tested for by fitting valves at the
tapping points; closing one or both valves should not affect the
instrument reading. If it does, one of the lines must be leaking.

Figure 5.3 illustrates some of these points. Items (g) and (h)
apply mainly to low speed gas flow. A correction may be required
if the lines are not all at the same temperature, so as to establish
the true pressures on which the velocity is based (see appendix 4).

In the laboratory it is often feasible to introduce flow
straighteners in the duct consisting of honeycomb or bundles of
cylindrical tubes, length-to-diameter ratios between 4 and 10 being
usual. These greatly reduce the errors due to insufficient settling
length. They are particularly helpful if the incoming flow has a
swirling (helical) motion.

When consulting the standards, the student will notice various
correction factors as well as various tolerances (degrees of uncertainty). The background to the corrections is roughly this: in

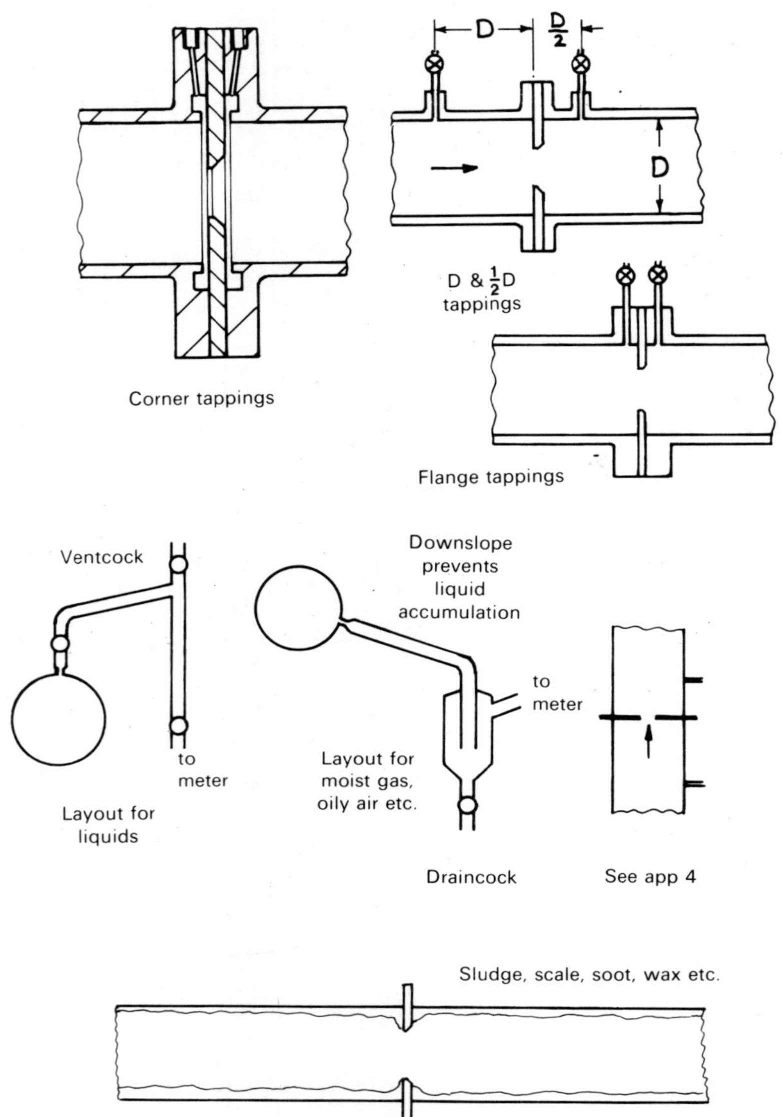

Figure 5·3

the constriction the flow is freshly speeded up, the boundary layer
is thin and the velocity profile almost uniform (plug flow). The
pressure drop measured refers to the increase of speed from approach
pipe to constriction. However, the velocity in the approach pipe is
not of plug flow distribution but depends on the Reynolds number, the
approach length and surface finish. The upstream pressure depends on

the central velocity, not the average; hence the need for correction
factors.

The orifice diameter should preferably not exceed two-thirds of
the pipe diameter. Although larger values give low losses, the
value of the reading is then unduly sensitive to the inevitable uncertainties of the approaching boundary layer thickness.

The settling length required upstream of the flowmeter depends on
how disturbed the flow is. If there are severe bends, globe valves,
etc., the official settling length may be up to 150 pipe diameters.
This is often impractical; with some loss of accuracy a much shorter
length is sufficient. [11, part 3]

In pulsating flow, the observed pressure on a gauge or manometer
will be the average pressure. In general, velocity will vary as the
square root of the pressure drop and unless the wave-form is known
and allowed for, the result will be quite wrong. There is a type
of flowmeter in which the pressure drop across a flame-trap element
is measured. This element has finely subdivided passages, widths
being 1 mm or less, so that viscous flow laws prevail. The pressure
drop is proportional to the velocity, therefore the average pressure
drop truly represents the average velocity. Some of these devices
suffer from relatively crude pressure tappings and should only be
used with the size of pipeline for which they are calibrated, otherwise the local speed at the pressure tapping is wrong (figure 5.4).

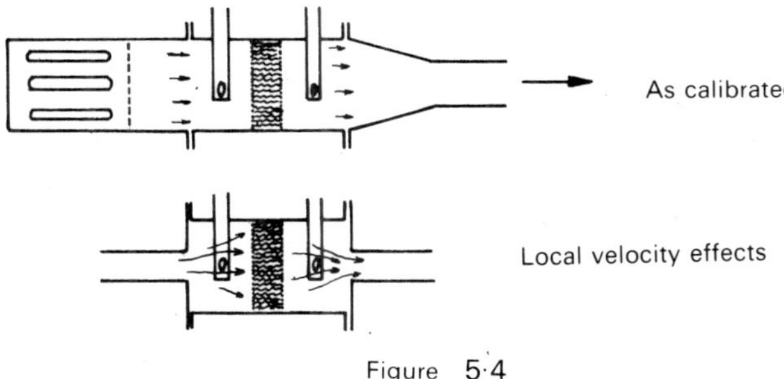

Figure 5·4

5.3 ROTAMETER-TYPE INSTRUMENTS

These operate at constant pressure drop, forming a self-adjusting
restrictor. The flow passes up a vertical tapered glass tube containing a body shaped like a fishing float or spinning top. The
constriction is between the outer edge of the float and the tube
wall. At zero flow the float rests on a support at the base of the
tube. At about one-tenth of maximum flow the pressure drop is
enough to lift the float into the graduated region. As the flow
increases, the float rises into the larger part of the tube to
establish a new equilibrium. The forces concerned are float weight
less its buoyancy in the fluid, the fluid pressure drop and any
viscous drag forces. The edge of the float has angled slots which

cause it to rotate. It also tends to stay central in the tube so
that static friction is avoided. Any instrument book will show an
illustration. The reading is normally taken at the widest part of
the float, where the constriction occurs.

These are excellent instruments provided they are correctly
applied, since the calibration is affected by fluid density, viscosity and thermal expansion of the float. The tubes are made of
low-expansion glass. Some floats are sharp-edged to reduce
viscosity-dependence but this makes them liable to wear prematurely
at the edge, by erosion and occasional contact with the tube.
Special floats are available for some fluids to give automatic compensation for temperature.

For flow rates too high for direct use, they could be used
between the tappings 1 and 2 of an orifice-type meter, but need
calibration. They are then subject to all the problems of orifice
meters with additional worries about part-blockage in the feed
lines.

5.4   TURBINE (PROPELLER) METERS

These consist of a propeller either driving a very light gear-train
or quite free-running, the movement being detected electrically or
by photocell.

There are two basic applications. In a pipe, such meters follow
the flow rate closely according to a stated calibration. The meters
normally include suitable flow-straighteners but assume a normal
flow profile. If the propeller is close to the pipe wall, the
fluid must be free from the commoner hazards of fragments of swarf
and scale, pipe jointing paste and particularly slivers of PTFE tape
used in pipe joints. The danger of creating such slivers is not
generally known in the pipe-fitting world (figure 5.5). Long fibres
of any kind may wrap themselves round the bearings and slow down the
rotor unbeknown to the user.

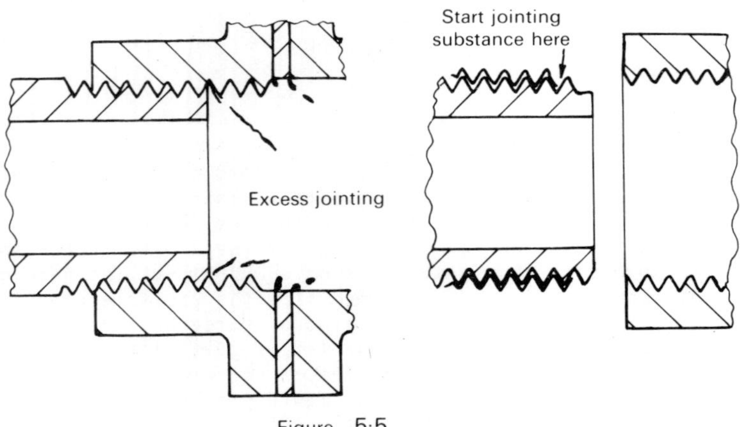

Figure 5·5

The other form of use is the anemometer or current-meter. In air
there is very little risk of unobserved wrapping-round of fibres;
in water it is appreciable. A further problem is that though the
rotor runs freely, its supporting structure impedes the flow; when
sampling the flow in a region, some flow is diverted thereby to
adjacent paths so that the flow through the meter is slightly
reduced from the free value. A particularly striking instance of
this occurs in rotating tanks sometimes used for calibrating flow
meters. The meter is immersed at some known radius and the speed
of the fluid is taken to be equal to the speed of the tank at that
radius. The flow meter, however, produces a drag on the relevant
annulus of fluid and the author has observed deficiencies up to 10
per cent (figure 5.6). The meter is to some extent measuring its
own wake velocity. The error may be estimated by observing floating
particles at the same radius and depth as the anemometer.

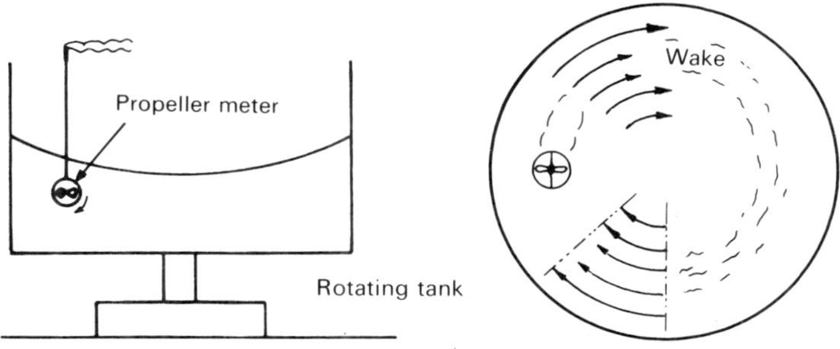

Figure 5·6

5.5   PITOT-STATIC TUBES

These measure separately the stagnation or total pressure and the
static pressure in a fluid, from which the velocity can be deduced.
They can produce slight speed-up of local flow, flow diversion and
reduction of total flow. Although the effect on flow measurement
as such may be negligible, it may be mentioned here that the wake
from the tube can have a considerable influence on any apparatus
downstream. To estimate these effects a subsidiary experiment is
required, using perhaps three tubes of decreasing size in the same
position (which is strictly speaking not possible for both the total
and static tapping positions at the same time). Small tubes are
slow in settling down owing to the very small passages - they tend
to vibrate, etc.; as far as possible they are used only in such
subsidiary experiments in order to validate the use of larger, more
convenient tubes. [13]

Although it is quite possible to make up one's own Pitot and
Pitot-static tubes, standard tubes have the benefit of declared
calibrations; at moderate velocities the calibration factor is 1.
A Pitot tube speeds up the flow downstream of the entry, reducing

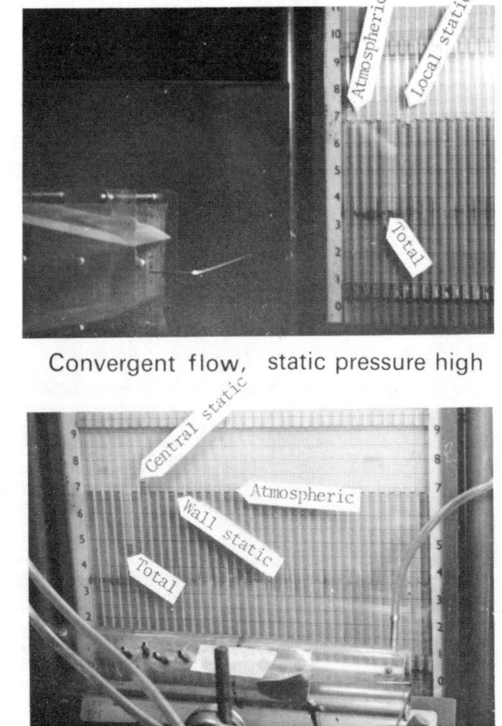

Convergent flow, static pressure high

Swirl: wall static high, central static low

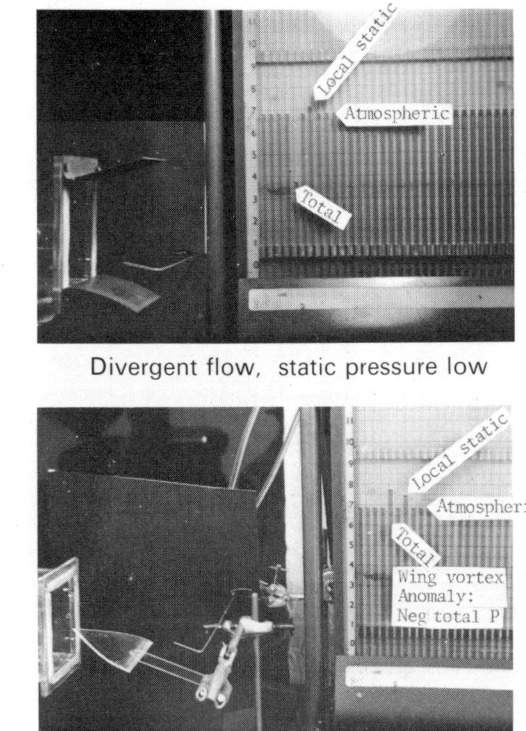

Divergent flow, static pressure low

Severe Vortex

Figure 5.7

48

the static pressure (nose error). The stem slows down the flow
slightly upstream of itself (stem error). In standard tubes these
errors cancel out if the tube is in a parallel flow region and is
small compared with the flow area. If it is of appreciable size,
the total pressure represents the approach speed but the static
pressure depends on the speed in the net space around the tube.
The mass-flow in the duct concerned is likely to be less than in
the same space before the tube was inserted, particularly if the
flow can divert readily to an alternative channel such as round the
other side of an aerofoil. It may be necessary to block the other
passages in the same proportion to obtain a meaningful measurement.

If the flow is not parallel, gross errors can result. The first
error comes from the fact that the nose and the side-holes are not
at the same station in the flow when the flow is convergent or
divergent. In swirling (vortex) flow the centrifugal effect produces an error. These are illustrated in figure 5.7. The Pitot
tube in these examples is in free air so that the nominal (free
stream) static pressure is atmospheric; it is evident that the
local static pressure is appreciably different.

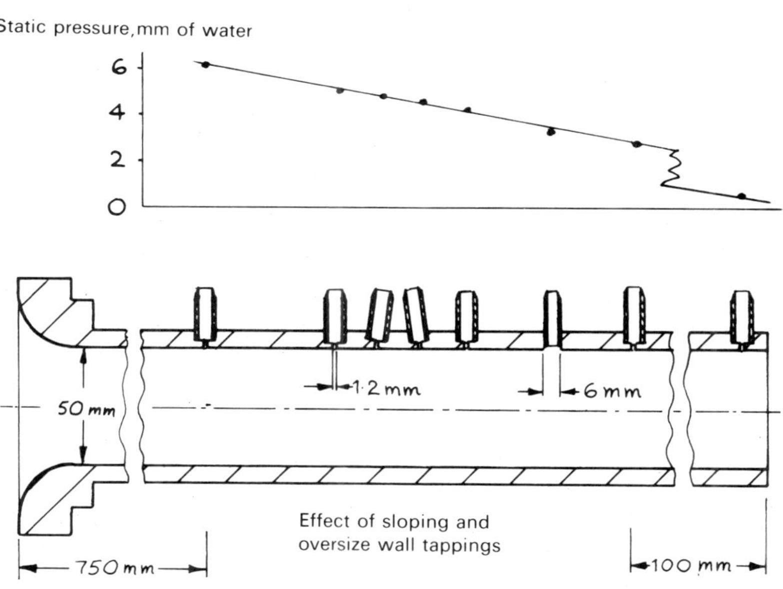

Figure 5·8 Wall tappings

The importance of correct wall tapping construction is shown in
figure 5.8. This is merely a demonstration. An accurate experiment
would require separate sets of apparatus to avoid suspicion of
mutual interaction.

In general a wall tapping must be free from burrs, normal to the wall, constant in diameter for a length of at least two diameters, and small enough not to disturb the flow. What is small enough depends chiefly on the boundary layer thickness. For the devices shown in figure 5.2 the sizes are laid down in the standards.

5.6 FLOW VISUALISATION

Fluid flow can be viewed and measured by inserting particles in the fluid. With water it is possible to generate bubbles by electrolysis, using stainless steel wire and a slight trace of electrolyte, usually nitric acid. The bubbles can be oxygen or hydrogen depending on polarity and can be started over limited parts of the wire by varnishing over most of the surface and leaving gaps where required. If the current is also interrupted cyclically, blocks of bubbles are produced. The best viewing is by side-lighting through a slit, against a dark background. This makes it easy to separate the views of particles at the edges of a stream from those at the centre.

The chief question is whether the particle follows the fluid correctly. In the low-speed regions of a system the settling speed (up or down) of bubbles or particles may be quite low since the force field acting on the difference between particle mass and displaced fluid mass is only 1g. In other places it may be much higher. Consider a flow of 25 m/s in a low-speed wind tunnel. If there is a cylinder of 10 mm diameter, the local flow may have a radius of curvature of 5 to 10 mm at various points around the leading edge. The local acceleration = (velocity)^2/radius = 25^2/0.005 or 25^2/0.0010 m/s^2 which is 12700g in the first case, or half this value in the second one. It may be questioned whether the path of the particle can still represent the flow of the fluid. Its velocity relative to the fluid will presumably be, in this example, 6000 to 12 000 times the low-speed settling velocity. This applies also to turbulence measurements using laser methods and particularly to intense vortices. The error goes up as the speed squared.

A second question arises from speed measurement from photographs. One way of obtaining the speed is to illuminate with a powerful stroboscopic lamp. A mercury vapour lamp gives interrupted light at *twice* the supply frequency. In using time exposure photography, whether with steady light or strobe light, the problem of focal plane shutters should be noted. A focal plane shutter takes a much longer time to traverse the frame than typical short exposure times, thus one end of the picture is taken later than the other. The nominal exposure time is varied by the width of the slit to give an equivalent amount of light on each part of the negative (see appendix 5).

QUESTIONS
(With comments and data)

These questions are intended to be worked from first principles using data provided in this book.

The Reynolds number is understood to refer to the small (throat) diameter, not the pipe diameter, in these questions; it is mean velocity × diameter ÷ kinematic viscosity, the viscosity being found from figure 5.9. The density of air may be taken as 1.293 kg/m^3 at 0 °C, 760 mm by mercury barometer, varying linearly with absolute pressure and inversely with absolute temperature. Take g as 9.81 m/s^2. 1 bar ≡ 750 mm mercury. 1 N/m^2 = 0.1021 mm of water at 20 °C. The change in pressure is calculated as $0.5(\rho_1 v_1^2 - \rho_2 v_2^2)$

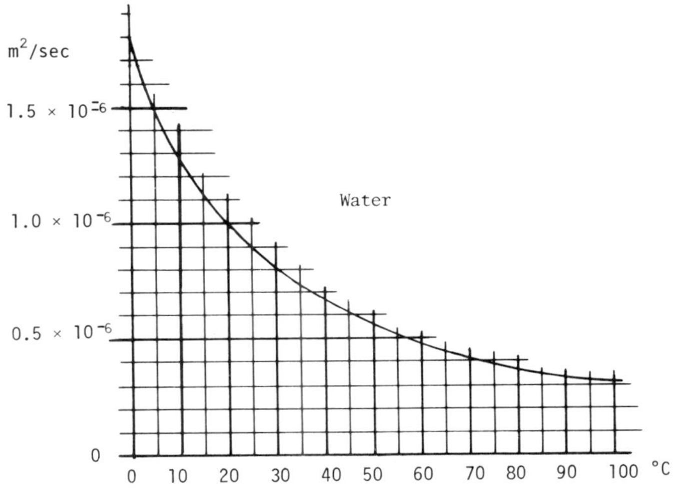

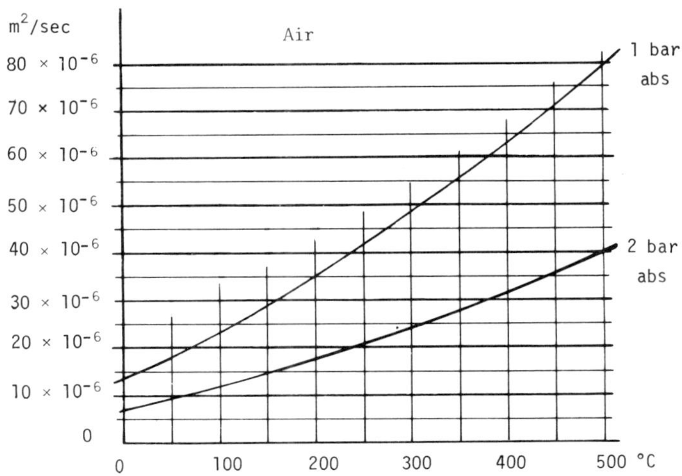

Figure 5·9 Kinematic viscosities of air and water

where ρ is the local fluid density, v is the local velocity.
C_d for a flowmeter is the ratio of real flow to flow assuming full velocity over the whole throat area. C_d is always < 1 owing to contraction of the fast stream and /or friction.

5.1 A smooth clean horizontal pipe of 150 mm internal diameter carries water at 60 °C. A venturi device as in figure 5.2 is installed, of 80 mm throat diameter. The pressure difference between 1 and 2 is found to be 0.2 bar (20 000 N/m^2). Find the flow in m^3/s and in kg/s. The density of water at 60 °C may be taken as 983 kg/m^3. Hint: Take C_d = 0.99 at first; find Re later, correct C_d if necessary.

5.2 A pipe 200 mm in diameter carries atmospheric air at a mean velocity of 20 m/s, temperature 20 °C.

(a) What reading would you expect to obtain on a water manometer connected to a small Pitot-static tube in the flow?

(b) What is the mass-flow of air in kg/s?

(c) If a flow-measuring nozzle of 100 mm diameter is installed and the mass flow remains unchanged, what reading should be obtained on a manometer connected between 1 and 2 if we ignore density change?

(d) What reading should we really expect with density change taken into account? (This part can only be answered by students with some knowledge of thermodynamics.)

5.3 A conical inlet according to BS 848 is connected to a pipe 0.5 m diameter. The reading on the manometer is 40 mm of water. What is the velocity and the mass-flow rate of air drawn in from an atmosphere at 45 °C, barometric pressure 735 mm? Take the air density in the pipe to be the same as the ingoing air.

Note that your working of questions 1 and 2 assumes that no further corrections are needed. The full procedure according to BS 1042 or similar standards includes some small corrections for approach pipe conditions.

6. HINTS ON BASIC MEASUREMENTS

6.1 MASS

Unless we are working in free space, mass is normally measured by
weighing. In cases requiring high accuracy, laboratory balances are
used, and to check on the equality of the two arms the load and the
known weights are interchanged. If any difference is found, the
average may be taken. The sensitivity is easily tested by adding
small weights to either side and noting the effect. The only con-
cealed error with this method is a change in the master weights
which may get heavier due to oxidation and general dirt, or lighter
due to wear and over-keen polishing.

The trickiest to weigh are hygroscopic substances, which attract
moisture, or volatile ones such as alcohol, hot water, camphor, etc.
The remedy is either to enclose them or to establish the mass-time
relationship in order to extrapolate back. This relationship may be
linear or may be slowed down by surface changes (skinning over, cooling, surface depletion, etc.). Spring balances may show slight errors
due to abnormal temperature. Platform scales may show slight variations according to where the load is placed. Friction errors may be
estimated from repeatability after disturbing the load, upwards and
downwards.

6.2 SIZE

Standard engineering methods usually produce satisfactory results in
measuring size. Errors may creep in with apparently regular bodies.
End faces of rods may slope slightly, opposite sides of rectangular
blocks may not be quite parallel or flat; plastics materials in
flat form vary considerably in thickness from place to place.

Cylindrical bodies may have ridges and hollows giving a constant
diameter yet be slightly non-circular. This may be detected by
resting them in vee-blocks and rotating them under a dial indicator
(figure 6.1).

Small round holes may be measured by passing short pieces of
known diameter through until finding one which fits tightly. A set
of mixed sewing needles is a useful source of small hard cylinders.
A needle and non-fluffy thread also serve to clean the hole before
measuring.

Measuring the volume of bodies with rough, cracked surfaces, also
porous and granular solids presents difficulties. Immersion in a
liquid and noting the increase of volume is subject to air-bubble

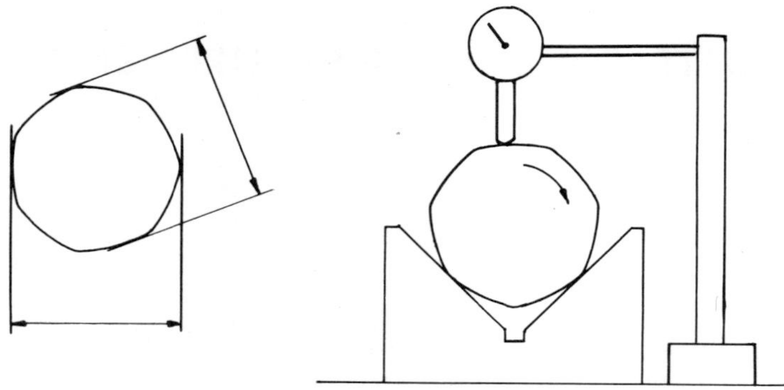

Figure 6·1

and absorption problems. If absorption troubles are absent, the
immersion method should work. It is useful to place a mirror in
the base of the container to view the underside. If bubbles cannot
be avoided, a vacuum process may be necessary. Suspend the body in
a vacuum chamber above a bath of liquid (vacuum oil would seem to be
the most suitable); when sufficiently outgassed, lower the body into
the liquid without releasing vacuum, or allow more liquid in, depending on available apparatus.

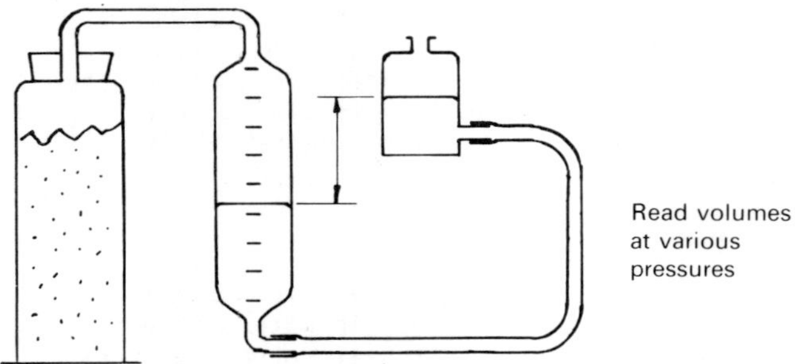

Figure 6·2

A method suggested for seeds uses Boyle's law, (seen at the
National Institute of Agricultural Engineering, Silsoe, Bedfordshire).
One way of setting this up is shown in figure 6.2. A vessel of known
volume with a ground-glass stopper or other positive closure is
partly filled with the specimen substance, connected to a graduated
vessel. A moveable reservoir bottle is used to raise the pressure.
After allowing time for the temperature to return to normal, the
readings of the pressure head and corresponding volume change enable
the net gas volume to be calculated. The internal capacity of
vessels which must not be contaminated may be measured similarly.

Another related method (for empty vessels), might be to use the
vessel as an acoustic resonator, in conjunction with a given small
pipe. The resonant frequency for a known vessel would be established

first; the resonant frequency of the vessel under test would indi-
cate the volume, frequency being proportional to $\sqrt{(\text{volume})}$, other
things being equal.

6.3 DENSITY

The density of liquids is usually found by floating hydrometers.
If greater accuracy is needed, bottles of known volume are obtainable; these are carefully filled and weighed. Alternatively, a
variation of the hydrometer method is to use a body of accurately
known size such as a large sphere, suspended with a fine wire in a
bath of the liquid and noting the change of weight.

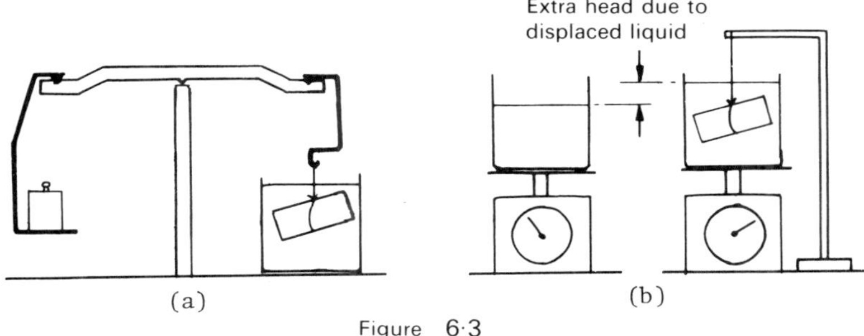

Figure 6·3

Two alternative layouts for this are shown in figure 6.3. These
may of course also be used to find the volume of the solid if the
density of the liquid is known. Layout (a) is obvious, the solid
is weighed in air and also when immersed in liquid. In layout (b),
the difference between the weighings is the weight of the displaced
liquid. This may be explained as follows. The net weight of the
body is obviously taken by the fixed stand. The extra force on the
bottom of the container is that due to the increased head of liquid;
it is exactly the same as if instead of immersing the body we had
added an *equal* extra volume of further liquid.

6.4 TIME

There are various accurate methods of measuring time; when measuring
the duration of some process the difficulty is in synchronising with
the timing device. Such end errors can be estimated by subsidiary
experiments. If a stop-watch is used, the operator must test himself
against known time intervals, such as a clock, over varying periods.
It is sometimes difficult to coordinate hand and eye; if several
observers are available, it is of interest to use one to give audible
signals to someone not watching the process while another person is
using the direct method, comparing the results.

If electrical or mechanical timing is used, there may still be
inherent delays, not necessarily equal at start and finish. The end
errors should be tested by varying the duration of the process. An

example is the measurement of gravitational acceleration by timing a
falling object. Repeating over many different heights of fall
should give enough data to find the time lags and perhaps also the
effect of air resistance.

A very convincing method of timing is based on photography. The
event may be photographed by time exposure with a rotating pointer
in the picture so that the effective exposure duration is externally
monitored. Focal plane shutters introduce an error because one side
of the picture is exposed later than the other, by a fraction of a
second; 'short exposures' are not short at all but are produced by
a narrow slit giving an equivalent amount of light.

A stroboscope light or strobe disc may be used, together with
time exposures.

One aspect of time measurement is speed of rotation. Students
are liable to overlook one form of error when using stroboscopic
methods. An intermittent light source is used to illuminate a spot
on a disc at a known frequency. When this frequency agrees with
the disc speed the spot appears stationary. It also appears stationary if the illumination frequency is 1/2, 1/3, etc., of the disc
speed, lighting the spot every second or third time. A better way
is to increase the illumination frequency until 2 or 3 stationary
images are seen, then the disc speed is 1/2 or 1/3 of the flashing
rate.

When a disc surface is not available but a shaft surface has to
be used, obviously image positions in the shadow are not effective.

A helical mark can be used. When the flashing rate is twice the
shaft speed, a double helix is seen, apparently stationary. At
lower frequencies, the same error as above can result.

When timing shaft speeds by photocell or by inductive methods, a
rough check of speed is important in case the counting circuit de-
tects the approach *and* the departure of a light-zone or notch.

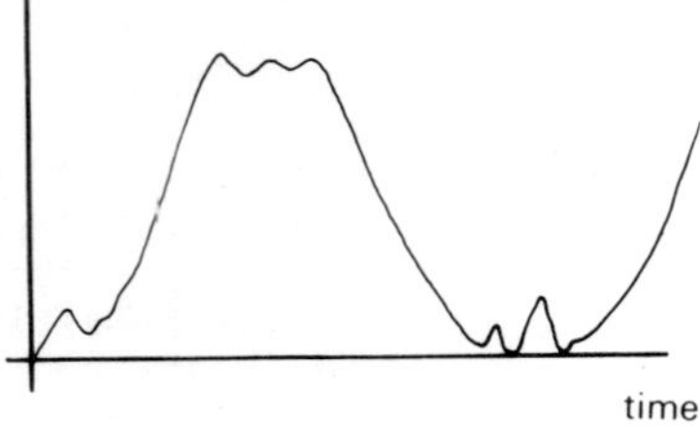

Figure 6·4

The photocell method is also liable to be confused by oily
patches which reflect light from supposedly dark areas. The wave-
form of the illumination could be as in figure 6.4; the circuit

56

for the count should be designed accordingly. One method is to
apply a half-way bias to the signal between the extremes and
arranging to detect the number of zero-crossings (in one direction
or both). Methods for doing this are outside the scope of this
book; the object of these remarks is to draw attention to the ease
with which spurious readings can be obtained.

6.5 STEADY PRESSURE

Steady pressures are measured by manometers or pressure gauges.
Normally they are read relative to atmospheric pressure. The observed reading is the pressure at the measuring point (tapping),
plus (usually minus) the gravity head of fluid in the connecting
pipe, plus any correction for the instrument. (The convention is
that if an instrument reads *low* it has a *negative* error and needs
a *positive* correction, and vice versa).

When measuring liquid pressures, errors can result from gas
bubbles in the pipe; when measuring gas, condensed liquid causes
errors. Transparent pipes reveal these effects and make it possible
to allow for them.

In some cases of high accuracy it may be necessary to observe the
temperature of the connecting pipe to find the density of the fluid
inside and correct the reading accordingly (appendix 4).

If the tapping point is in a region of high speed flow, swirling
flow, if it faces upstream or downstream or if it is distorted by
burred edges, the significance of the pressure needs to be given
due consideration.

Manometers suffer mainly from surface tension errors. If the
pipes and liquid are uniformly clean, the effect is constant provided the pipe size is constant. In vertical pipes of 6 mm diameter
or larger, the meniscus is usually fairly flat in the centre. A
visual extrapolation to an infinite diameter allows us to estimate
the remaining error. In inclined manometers and in two-liquid differential manometers such estimates are not so easy; high cleanli-
ness and good wetting are important. A little wetting agent added
to water or a small amount of solvent added to oil or paraffin can
help in giving consistent contact. If small pipes have to be used,
the amount of capillary rise to be expected may be found in textbooks on physics. In engineering experiments a zero-check is
usually sufficient.

When a liquid column manometer is inconvenient, pressure gauges
are usually used. The commonest form uses a curved flattened tube
(Bourdon tube). This form of tube uncurls slightly under internal
pressure, curling up under suction. The movement is magnified by
suitable mechanisms such as levers, gear quadrants or cam action.
Such tubes are affected somewhat by temperature and very considerably by overpressurising, which deforms the curled pipe, changing
its zero and in severe cases totally altering its elastic deflections. No laboratory should be considered fully equipped without
a gauge tester. A tester consists of a pump feeding the gauge and

at the same time feeding a highly accurate piston of known size
free to slide vertically in a cylinder and carrying a selection of
weights. The cylinder and weights are rotated by hand during testing to free the vertical motion from static friction.

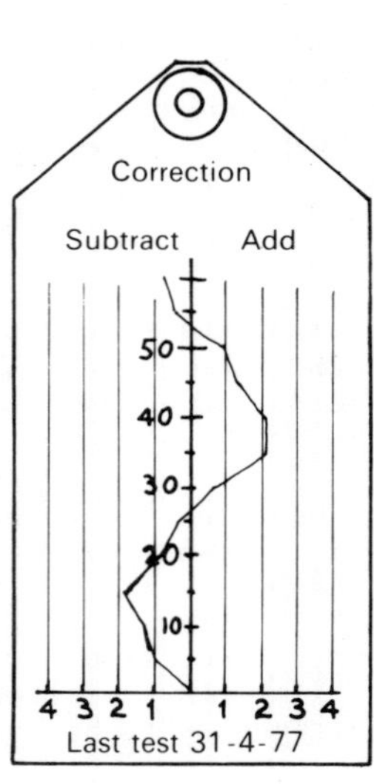

Figure 6·5

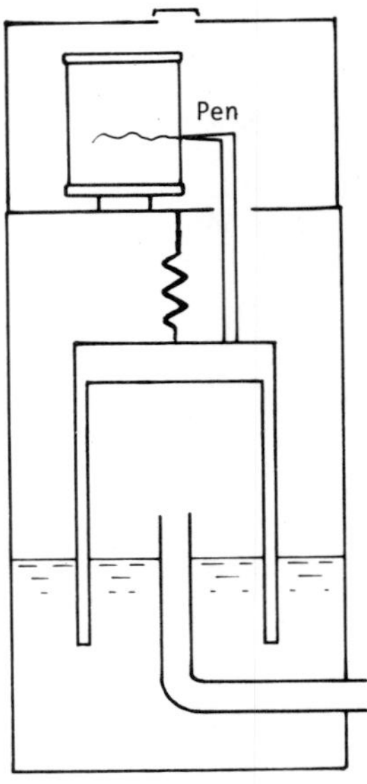

Figure 6·6

The correction to be applied may be presented as in figure 6.5,
in the form of a label tied to the gauge. It is usual to present
it not as the error but as the + or - correction, to be added to
the reading.

For low pressure differences from atmospheric, of the order of a
few millimetres of water column, a floating bell instrument may be
preferred (figure 6.6).

The bell has ballast on the bottom and may be suspended by a
very soft spring. There is usually a guide rod. The pressure or
suction is fed to the inside of the bell and acts on a relatively
large area. Thus the forces concerned come from the pressure of
the gas, the buoyancy of the immersed parts and the spring. These
forces are made relatively large to overcome surface tension and
friction effects. The surface tension force varies only as the
diameter, pressure forces as diameter squared.

The sensitivity can be very high since for an increase of height
the chief change of force can be the reduced buoyancy of the side-

walls of the bell. Such gauges can easily be made recording. Once calibrated, they should give very little trouble unless the spring gets tired. The disadvantage is the large volume change which makes them slow in action when long pipes and small pressure tappings are used. Instruments using diaphragms whose deflection is detected electrically have much smaller volume changes but depend on calibration by a primary instrument.

Manometers and diaphragm instruments are ideally suited for measuring pressure differences for flow measurement. Manometers can be made recording using a float and lever, generally with mercury as the fluid. Since they are usually required for a pressure difference, both surfaces are sealed from atmosphere; the float rests on one of the surfaces and turns a shaft which passes through a low-friction seal to atmosphere where it operates a pointer or pen. Some degree of checking and maintenance is needed to keep down friction and leakage at this seal; there are some leakproof instruments with magnetic transmission of the movement. The author prefers the positive movement which can be checked for friction directly.

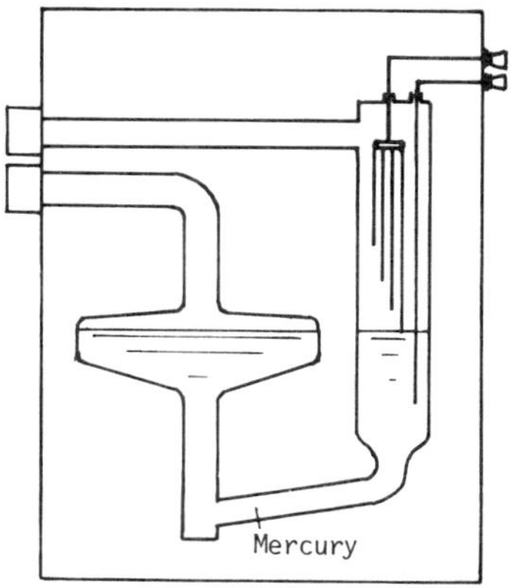

Figure 6·7

An alternative form of read-out free from this problem is to immerse a series of carbon resistance rods in the mercury (figure 6.7). The resistance varies with the exposed length of rod. In theory one rod would be sufficient but it is easier to obtain a linear scale or square-root scale by using several rods. The electrical leads are easy to seal against leakage. Obviously the electrical signal needs calibrating. There is a possibility of error due to scum on the mercury; unfortunately it is not usual to make the chambers transparent.

Many instruments are unsuitable for direct use with water but can be used to measure water pressures and velocities by using a small feed of air into the lines. In Pitot tubes there are appreciable errors at the point where the excess air bubbles out and modifies the flow. In horizontal orifice meters with corner taps no great error should occur, since the bubbles emerge into low-speed zones. If the exits are at differing heights, see appendix 4.

Vacuum gauges for sub-atmospheric pressures are perhaps better called suction gauges, as long as they use the same principles as pressure gauges. When the pressure is low enough to qualify as high vacuum, measuring relative to atmospheric pressure becomes inaccurate and we resort either to manual instruments that take a sample of the gas concerned and compress it to perhaps 0.001 of its former volume and then measure the pressure, which may involve condensing some vapours, or we resort to indirect vacuum gauges using the principle of heat loss from an electrically heated filament, ionisation or related electrical effects. Some of these instruments are made selectively sensitive to a gas not common in air such as helium, Freon refrigerant, etc., for use as leak detectors in vacuum apparatus. Any textbook on vacuum technology should provide further details. The principle of leak detection is this: we connect the apparatus to be tested to a vacuum pump via the leak detector (the two generally come as a complete outfit); then we squirt small amounts of the tracer gas locally at various parts of the apparatus; when the tracer gas is drawn in through the leak, the detector gives an audible signal almost immediately.

6.6 PULSATING PRESSURES

Pressure gauges and manometers have too much inertia to follow rapid changes; the devices used for rapidly changing pressures use diaphragms whose deflections are detected electrically by a change of capacitance or inductance, or by piezo-electric means using a suitable crystal. For extremely high pressures, the resistance change of manganin wire under hydrostatic pressure is used.

Devices that translate the pressure signal into another form come under the general heading of pressure transducers. These are obtainable from various instrument makers and come ready calibrated. In addition one may calibrate the complete train of transducer, intermediate devices and read-out as a whole. Whether a static calibration is relevant under rapidly changing conditions needs to be examined. Piezo-electric crystals give the greatest effect when pressure change is rapid; they can only give out a very limited amount of energy. During slow changes they lose charge easily, therefore to record a slower change correctly, very high impedance is needed in the instrument circuit. If the impedance is insufficient, the reading tends to be proportional to rate of pressure change, giving a jerk-meter which is in itself a useful instrument.

In capacitance or induction-based transducers connected to suitable amplifying circuits there should be very little doubt about the meter following the local pressure at the diaphragm provided that spurious electrical fields do not impose extra voltages on the leads. With capacitance transducers the line capacitance may become appreciable; short cable runs and low-capacitance cables should

help but wires spaced far apart to cut down mutual capacitance are
then in danger of forming a loop aerial picking up spurious electromagnetic influences.

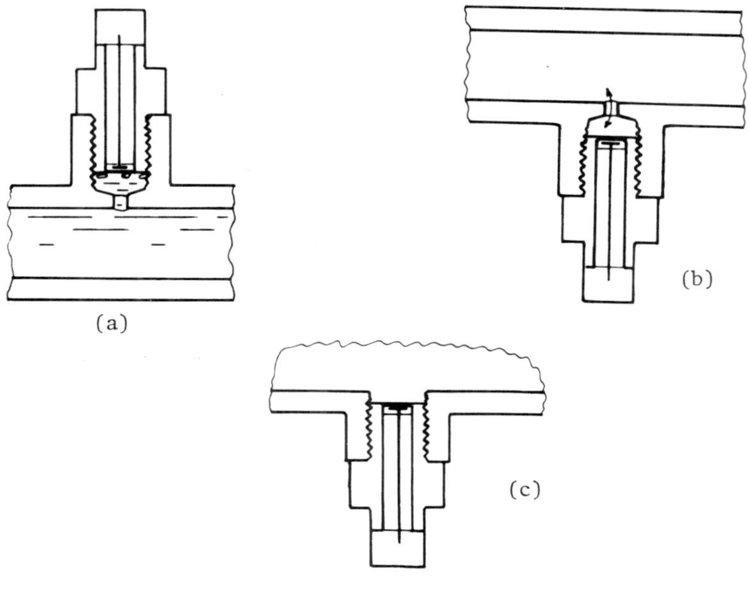

Figure 6·8

   The electrical side is unlikely to be the worst source of error:
consider a transducer installed in a liquid-filled apparatus as
shown in figure 6.8a. Air bubbles are likely to accumulate under
the diaphragm during operation; this softens the whole system,
modifying its dynamics. It is safer to install the transducer from
below. However, if installed as in figure 6.8b, the dynamics are
still falsified since the large chamber with the deflecting diaphragm connected to it is in series with the small access hole,
forming a cavity resonator of a certain frequency which may inter-
act with the system. Preferably the installation should be as in
figure 6.8c. Alternatively, pressures may be obtained by applying
strain-gauges to an existing pipe already part of the system.

   In gases, the orientation is not important but the cavity effect
should still be avoided.

6.7 FORCES

Forces are one of the most basic features in engineering measurements; they are considered at the end of this chapter because they
form an ideal launching point for discussing inherent errors in
structural experiments.

The traditional force meters are weighing machines, spring
balances and proving rings, all readily calibrated by weights or by
testing machines based on weights. For remote reading and for
fluctuating loads it is more convenient to use commercial ready
calibrated load cells, chiefly those based on strain-gauges. Those
still have some deflections and may therefore falsify the behaviour
of the structure as explained below. The stiffest load cells use a
quartz piezo crystal (but see section 6.6). The way of ensuring
minimum interference is to put strain-gauges on the structure itself,
on all the relevant members.

The importance of possible errors due to deflection may be illustrated by an extreme case. Supposing we wish to find the weight of
a four-legged table by placing one leg on a weighing machine, the
other three on rigid blocks of the same height as the *unladen*
weighing machine. If the table is perfectly rigid, the slightest
deflection of the weighing platform will cause it to read zero. If
we jack it up, the reading can be up to half the table weight. This
is an example of load-shift between redundant supports which cannot
always be avoided; it is important to learn to deal with it.

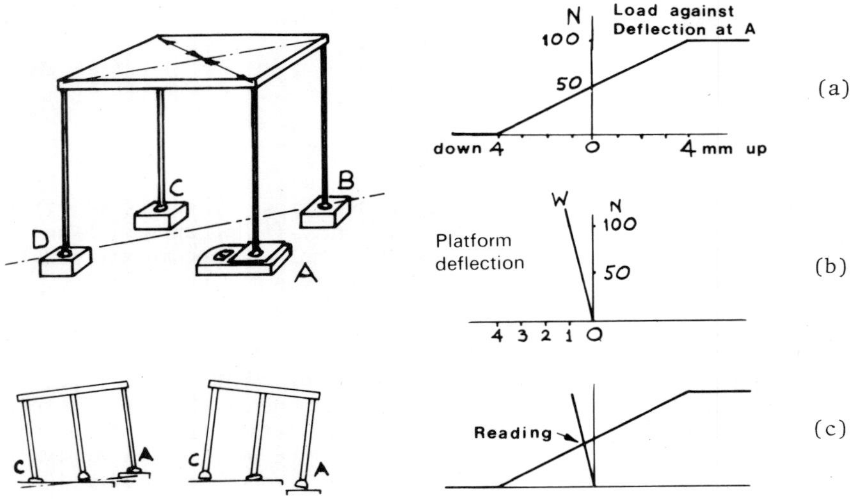

Figure 6·9

As a simple example, take a table with four legs in a 1 metre
square configuration (figure 6.9). Assume the table to be completely
rigid but let each leg include a little rubber pad of stiffness 50
N/mm; the table weighs 20.38 kg (200 N) and one foot is placed on a
weighing machine, the other three feet on rigid blocks at the same

height as the weighing machine. If the weighing machine did not deflect at all, it would obviously read 1/4 of the total weight, 50 N.
For very small deflections the resilience of the pads would ensure
that the reading is not far from 50 N. Our object is to establish
what the reading would be for weighing machines of various stiffnesses; thus we wish to establish a law relating the force in a
leg to the displacement beneath it. If the system is quite linear
we can take the easy way out of finding the displacement when leg A
is hanging free, under zero load. In this condition, taking moments
about the two nearest legs B and D shows that the force in the most
distant leg C is also zero, thus the compression of pads A and C is
zero. The pads B and D support the whole load, 100 N each, and
therefore compress 100/50 = 2 mm, the table tilts slightly and the
free leg must finish at least 4 mm below the original level. There
is an additional effect due to the angle of tilt. The angle of tilt
= 2/0.707 or 4/1.414 radians. The c.g., at some distance h above
the base, is displaced laterally towards leg A by 2 × h/0.707 mm,
giving an extra moment about the diagonal X-X. However, the rubber
pads on B and D are tilted by this action and for the present
example we assume that the restoring moment due to the tilt just
cancels that extra moment, so we can ignore this effect in the present example.

Thus we can now say that the leg A will have a load of 50 N if
not displaced at all below the common level of the other supports,
varying linearly to zero when deflected 4 mm downwards.

The argument can be reversed if the base of leg A is forced upwards; the plot of load versus displacement is shown in figure
6.9a. The highest load possible on leg A is *half* the table weight;
B and D have zero load, A and C are both under 100 N loads and
therefore shortened by 2 mm, thus demanding a step height of 4 mm.

Now consider the leg A not free but standing on a weighing
machine of such stiffness that the platform goes down 1 mm for 100
N. If it is set up carefully to be just at zero load when at the
same height as the packing blocks under feet B, C, D, we can draw
its operating line, O-W, on the same graph. Where the two
lines intersect, we find the reading to be expected. (If the
set-up is not exact the operating line would simply start from a
different point, still having the same slope, the platform being
pushed downwards as the force increases.)

6.8 ABOUT STRAIN-GAUGES

Strain-gauges are used to detect extension or compression by changes
in their electrical resistance. A component under tensile strain
lengthens and also gets slimmer in cross-section according to its
Poisson's ratio. This alone would affect its electrical resistance
but there is a further effect due to the material itself.

If the original component had a length L, width b and thickness
t, its resistance R was resistivity times L/bt.

When extended by dL, the width and thickness reduce according to

$$db/b = dt/t = - \nu \, dL/L \quad \text{(note the minus sign)}$$

If the resistivity did not change, dR/R would be $(1 + 2\nu)dL/L$.
However, the materials used also change resistivity, increasing
under tension and decreasing under compression. The increased
effect is given by the gauge factor. Seeing that ν varies from
material to material, the manufacturer normally expresses the sensitivity of the gauge in some unambiguous form, for example:
$R = 10 \, \Omega$, $dR/R = 2.05 \, dL/L$, or some similar statement.

   This short section cannot hope to do justice to the whole topic
of strain-gauges; manufacturers' handbooks and a recent textbook
should be consulted. However, a few commoner points are discussed
here.

(a) The gauge adds to the stiffness of the component (reinforcement),
    which may be significant in very thin components. The stiffness
    of the gauge is stated in the catalogue.
(b) When used in bending, the gauge element is further from the neu-
    tral axis than the component's original surface, so a correction
    needs to be applied.
(c) The adhesive may creep with time. This may be detected by
    checking the zero before loading and after releasing the load.
(d) The electrical instrumentation, a form of Wheatstone bridge, may
    produce a zero shift due to heating of its own components or due
    to electrical energy fed into the gauge element.
(e) If the component undergoes thermal expansion differing from ther-
    mal expansion of the gauge, the gauge will be under strain even
    if the component is not under load. Older gauges do not allow
    for this and have to be used with a dummy gauge subject hopefully
    to the same thermal conditions as the working gauge which is very
    difficult to ensure. Modern gauges are temperature-compensated
    for a particular substrate. This is achieved by making the gauge
    of a material whose temperature coefficient of resistance just
    compensates for the resistance change due to the differential
    thermal expansion. This is only possible over a limited range
    of temperature.
(f) If a gauge is used on a curved surface there may be some doubt
    about its reading, especially if a convex surface is under com-
    pression or a concave surface under tension, the gauge tending
    to lift away from the surface. This trouble should show itself
    by lack of repeatability and non-linear behaviour.
(g) A gauge is basically intended for tension or compression along
    one axis (though gauges are available grouped together in
    rosettes, etc.). A single gauge will show a small resistance
    change if strained transversely; this is additive to its normal
    behaviour, (cross-sensitivity). In modern foil gauges or electro-
    deposited or vapour-formed gauges it is possible to make the
    transverse links broader than the sensitive lengthways threads
    so that they contribute only a small part of the total resistance;
    therefore any changes in this resistance due to cross-strains
    only have a small effect on the principal reading. The cross-
    sensitivity of a gauge is stated by the supplier.
(h) A method of detecting bubbles in the glue has already been men-
    tioned in section 4.1.

(i) Resistance changes and other side-effects in the leads may be
     largely eliminated by running an additional pair of dummy leads
     alongside the main leads up to the gauge position. These are
     securely shorted together near the gauge and are connected into
     the adjacent bridge arm which is of the same basic resistance as
     the gauge. Any effect adds an equal resistance or equal spurious
     voltage to each arm so the balance is practically unaffected.
     Another system using three wires in effect extends the galvano-
     meter lead up to the gauge itself so that the incoming lead is
     part of the gauge arm and the return lead part of the next arm.
     This seems less desirable since it greatly extends the galvano-
     meter circuit, thus exposing the sensitive imbalance reading to
     stray influences picked up by the long lead, particularly if
     alternating current is being used or if the loading is fluctua-
     ing.

QUESTION

6.1 A platform similar to that in figure 6.9 has feet spaced in a
square 1.5 m × 1.5 m and weighs 10 t. Each leg is fitted with a
freely tilting foot-piece of compressive stiffness 1 t/mm. The legs
may be regarded as rigid. Leg A stands on an adjustable base, legs
B, C, D on fixed rigid blocks of equal height.

   Find the relationship between force and displacement at the adjustable base as a graph something like figure 6.9. Hence find the
reading given by a weighing machine placed under leg A if its stiffness is 5 t/mm and it is placed when unladen (a) at the same height
as the fixed blocks, (b) 0.5 mm higher, (c) 1 mm lower, ignoring
the effect of centre-of-gravity displacement laterally.

The platform dimensions are not relevant, they are given merely
as an aid to visualisation.

7. TEST RIG PROBLEMS

In the following examples the student will notice three discernible
types of difficulty: redundant support problems as in chapter 6,
errors from friction and from load-line displacements.

7.1 CUTTING TOOL EXAMPLE

Consider an experiment to measure cutting tool forces. Direct measurement of tool deflection is not entirely satisfactory because of
bending due to temperature gradients. Strain-gauges on the stiff
tool-shank are inappropriate at the low stresses because they would
produce very small signals. Perhaps the best way is to add a load
detector under the tool. This of course increases the stiffness
thus altering the dynamics but usually there is no objection to
this in machining tests (figure 7.1).

In figure 7.1a we have a weigh-beam transmitting force to a remote load cell. The whole arrangement is easy to calibrate by hanging weights from the tool tip, thus taking care of the multiple support effect. Friction can be estimated and would in practice be
minimised by greasing, thus giving a satisfactory *calibration*.

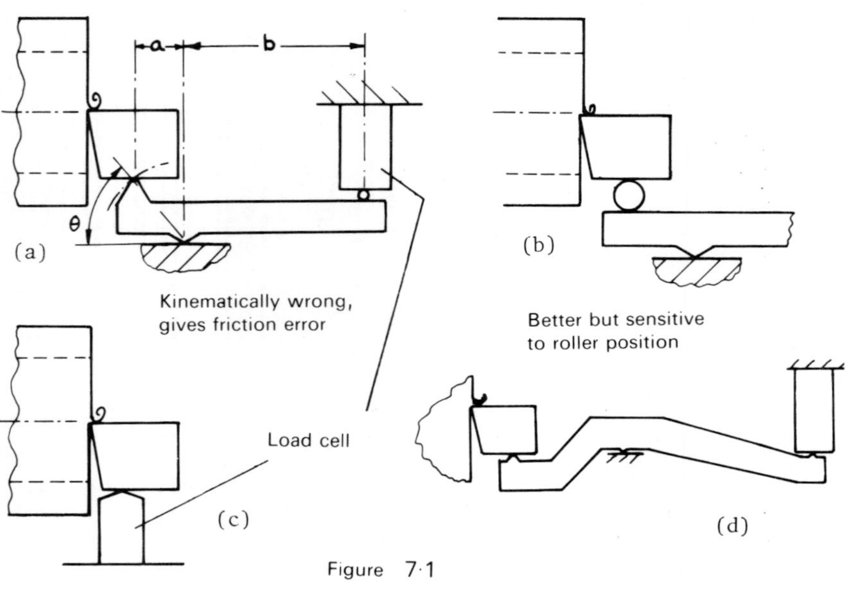

Figure 7·1

However, the lubricant will gradually disappear due to heat and movement, thus friction forces will gradually become important some time after calibration, spoiling the accuracy.

At increasing loads, the friction force will oppose the load so that a reduced reading results. If $\mu = 0.3$ and the angle $\theta = 45°$, the friction contribution is readily found by taking moments about the knife-edge; the load cell reading is 30 per cent lower than a true frictionless reading.

A better arrangement interposes a roller, but with the danger that during vibration the roller may get displaced (figure 7.1b). Arranging a stiff load-cell under the toll (figure 7.1c) is the safest but not altogether safe: the force measurement cannot be totally divorced from tool deflection, thus bending caused by thermal expansion may be significant since the top surface of the tool shank will be warmer than the lower surface, due to chip friction. To monitor this effect one should check whether a zero reading is obtained *immediately* the cutting stops. A kinematically 'correct' equivalent of set-up 7.1a is shown in figure 7.1d. The line of action is at 90° to the line joining the three pivots so that the frictional components have the least possible moment about the fixed pivot.

7.2 BEAM EXAMPLE

Figure 7.2 shows some bending tests. Set-up (a) is subject to a friction error similar to that in figure 7.1. Here however we have rather more deflection. As the beam deflects, a increases, b decreases, so we also have a perceptible geometry change.

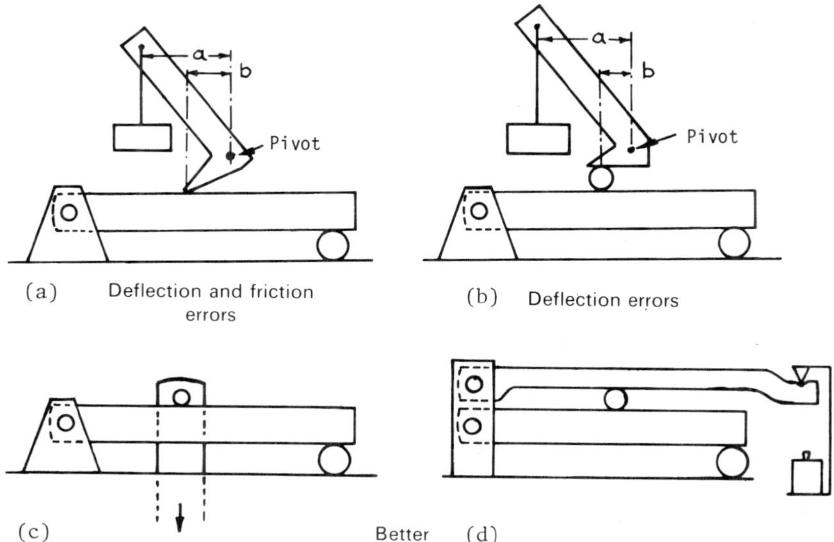

Figure 7·2

Set-up (b) is somewhat better because sliding friction is removed.
The consequences of deflection are still there, a increasing while b
decreases (at a reduced rate).

In layouts (c) and (d) we have true vertical movement when the
roller is at mid-span. At all other positions there is still a
small error due to roller movement towards the centre, usually negligible.

7.3 BEARING EXAMPLES

The need to consider displacement effects is brought out again in
figure 7.3 which shows a demonstration version of a journal bearing
comprising a shaft, a half-shell and a loading arrangement rigidly
connected to the shell. When the shaft is rotated, the friction
force swings the load system through a small angle. The bearing is
of larger radius than the shaft, to encourage wedge-film action:
this is shown greatly exaggerated. The bearing takes up an eccen-
tric attitude so that its centre is displaced by an amount δ. If
we ignored this, the torque to be overcome would seem to be $W\ell\theta$,
giving a friction coefficient at the shaft surface (tangential force
$\div$ normal force) of $\ell\theta/r$. The true torque however is $W(\ell\theta + \delta)$. It
is quite possible for δ to be perhaps $0.0003r$ and the real μ to be
0.002; thus ignoring the shifting-over gives a 15 per cent error.

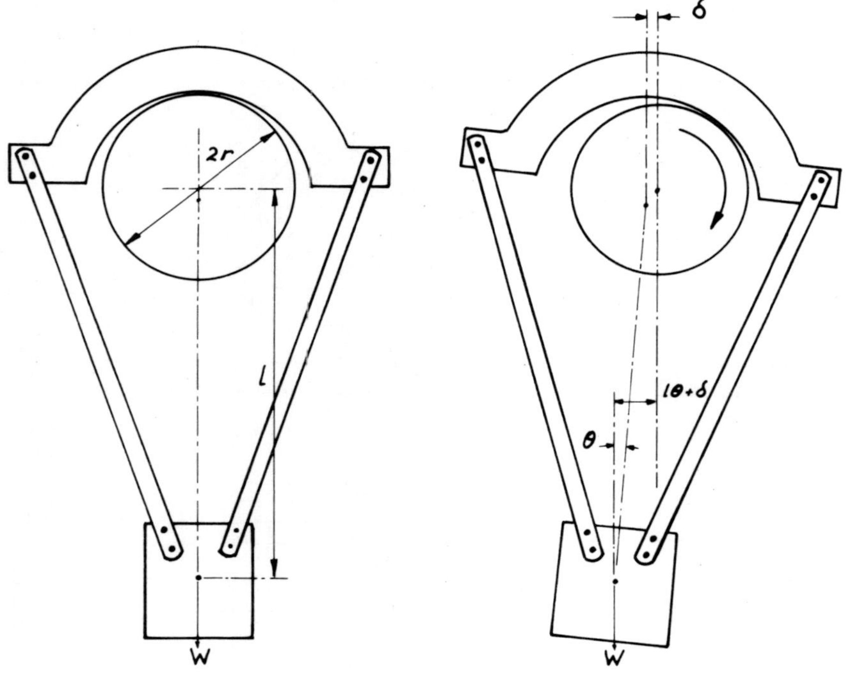

Figure 7·3

68

b
a
C
W
F
x₁
μN
R
Ø
N
W
(a)
F
R
μN
N
W
(b)
μN
F
R
N
W
(c)
Microscope
R
N
W
(d)
Adjustable

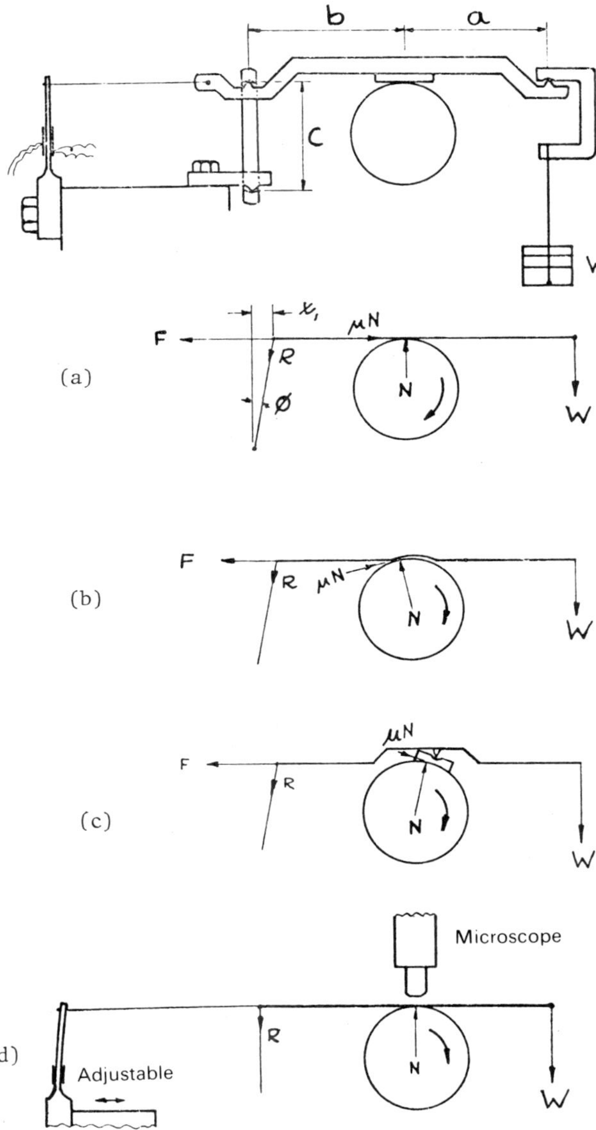

Figure 7·4

If we reverse rotation, the result is just the same, still concealing part of the effect. To obtain the true effect one may measure
the displacement of the c.g. of the loading system directly, perhaps
by a microscope with an eye-piece scale.

Another friction apparatus is shown in figure 7.4. Here the problem is apparently avoided by measuring friction force directly,
using a strain-gauged cantilever previously calibrated by weights.
The little problems involved in this are discussed later. For the
moment we assume that we obtain a true measurement of the force F.
The load points are collinear so that small *angle* changes do not
affect the leverage. The errors in this apparatus are considerably
more subtle than the preceding.

Consider case (a) in which the contact remains flat as the shaft
begins to rotate clockwise. As the load-meter deflects, a small
displacement x occurs. The down-load (equal to the vertical reaction N) changes from the nominal $W(a + b)/b$ to $W(a + b)/(b - x)$.
The horizontal force at the contact point $= \mu N$. Resolving horizon-
tally, the measured force $F = \mu N - Rx/c$.

Provided that ϕ is small, we may take $R = N - W$, therefore

$$F = \mu W(a + b)/(b - x) - [W(a + b)/(b - x) - W] \; x/c$$

$$= \frac{W}{b - x} [\mu(a + b) - (a + x)x/c]$$

The student can readily check that if $c = ab/\mu(a + b)$ the system
is self-correcting for small displacements x.

However, soon the pad will wear to shape (b); now the contact
moves to the left, the normal reaction is no longer vertical and
tends to reduce the reading F. In the limit, the groove will be so
deep that the friction angle opposes the slope angle at entry. Then
the beam and pad can slide downwards to the left, the reading F can
fall to zero indicating zero friction which is manifestly quite un-
true.

To overcome this error we can make the pad self-aligning (c), its
pivot kinematically correct on the horizontal line of action. Now
N has a component to the right which augments the friction force and
we still have a wrong reading.

The best answer is to ensure that horizontal movement is cancelled
out by displacing the load-meter (d).

The problem of case (b) can now be tested for. If small but perceptible movements to left and right can be made with no change in
the measured force, then we are free from groove effect.

7.4 CALIBRATION EXAMPLE

A convenient load-cell giving electrical signals from small loads is
the strain-gauged cantilever. It is readily made up as required and

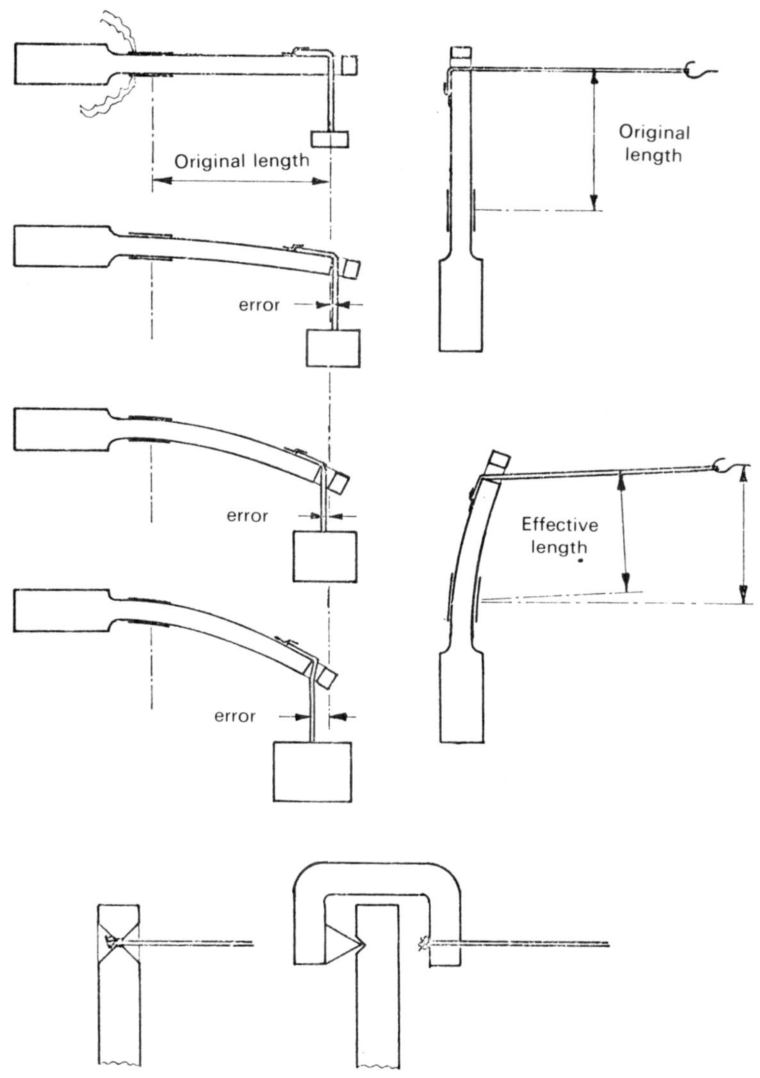

Figure 7·5

easy to calibrate with weights. To get reliably large resistance changes in the strain-gauges, appreciable stresses are involved; this implies a thin member which in turn gives appreciable deflec- tions. Though these do not give large errors, the student should not be tempted to ignore them, especially since the effect is non- linear. Figure 7.5 shows the problem clearly. The strain-gauges are normally applied to both faces of the cantilever and wired into adjacent bridge arms, making the resistance changes additive for

71

bending but insensitive to the direct tension component. The reading will generally be proportional to the bending moment. The dia-
grams make it quite apparent how the bending moment is affected by
the deflection during calibration and during an application to horizontal force measurement.

A convincing way of measuring the true offsets is by photography
with an accurate graph paper placed in the picture. At the risk of
labouring the obvious it must be said that a graph paper in the
background appears smaller than an object in front of it. The
easiest and most positive way of establishing the scale factor, is
also to place a small strip of graph paper or any other marker of
stated size in the plane of the object, for example, stuck to the
string.

This type of error is often small and will chiefly be important
when comparing data from several experiments.

A comment is required on the use of string, etc. With dead-
weight loading, extension of the string is not important. In any
other case there could be a genuine problem here. Wire may be too
stiff to give true values, string extends with time, is affected by
temperature and moisture. Perhaps fine jeweller's chain is the
most appropriate, as used in some laboratory balances.

A simple way of measuring horizontal forces is the pendulum
method. A known weight is suspended from a string of known length.
The horizontal force is applied at the junction of weight to string.
The horizontal deflection from the zero load position is measured,
giving the horizontal force by simply resolving the horizontal and
vertical components of the force in the string.

7.5 DISTRIBUTED LOADS

Such loads must be applied with considerable thought if we are to
avoid large friction errors. Aircraft structures are test-loaded
by large numbers of point-loads, using bags of lead shot (if sand
is used, there may be a wetness problem), or by complicated lever
systems that divide the load in the required proportions.

In the laboratory, a uniformly distributed load can be produced
by one layer of rollers. Anything more than one layer stiffens up
the structure and gives entirely misleading results.
This makes it very difficult to measure loads due to granular
solids, important in data for earth-retaining walls, cement silos,
etc. It is well established that the pressure distribution on such
walls depends strongly on the elastic deflections. The granular
material is relatively stiff until it starts to slip; it is liable
to reduce the load where deflections are largest and transfer it to
stiffer parts. This means that it acts as a supplementary beam;
indeed it is capable of arching over small openings completely.

The most realistic way of modelling such structures is one which
simulates the elastic strains in the granular material and in the
confining walls in the same proportions as the real-life structure.

This, however, is not convenient for analysing pressure distribution; local pressures are more convenient . To obtain a local pressure, a part of the wall is replaced by a diaphragm with strain-gauges or other detecting means. Truly to represent the piece of wall taken away, the gauge should have no more deflection than the wall.

We could allow the diaphragm to deflect and then force it back; the difference between force on deflection and on restoration would be very large, giving little guidance about the true pressure value.

An alternative is to establish artificially the pressure at zero deflection by using a series of gauges of varying stiffness and constant diameter, extrapolating back to the unattainable case of zero deflection (that is, equal only to that of the piece of wall taken away). It would be unwise to vary the stiffness by varying the diameter, this could give a series which misses the real point, by allowing the shear strain in the granular bed to be the same at all readings. The point is to ensure that we can get back to the value for a wall with no hole in it, where the shear strain in the bed will be very small.

This illustrates a general point. In an experimental series we try to vary one *fundamental* parameter at a time. When we change one external feature we may find we change several fundamentals unless the feature is chosen critically.

7.6 FORCE-FREE DEFLECTION MEASUREMENT

In light model structures it is important to ensure that deflection measurements do not disturb the condition. Workshop dial indicators exert appreciable forces. There are various convenient force-free methods. The travelling microscope is accurate but slow. When using vernier gauges or micrometers, electrical contacts or feeler gauges of thin paper can be used to detect the beginning of contact before appreciable forces arise. If a pneumatic gauging device is available, this forms a quick and convenient means. The author has used this successfully in surveying the vertical deflections of a disc (figure 7.6). A normal workshop height-gauge was fitted with a small nozzle connected to a Solex pneumatic gauge (shown schematically). When the nozzle is well away from a surface, the air can escape freely and only a low reading shows on the manometer. As the nozzle is brought close to the surface, back-pressure is built up. By using a small nozzle, ~ 0.5 mm diameter, the force exerted on the structure is minimal, clearances of the order of 0.02 mm can be maintained consistently and the method is not sensitive to small angles of slope. At larger slopes, the nozzle is tilted slightly and a fresh reference level is established, using some fixed point on the structure.

If a Solex gauge is not available, an equivalent is easily improvised from laboratory ware and needle valves. The essentials are a filtered air supply, a bubble-tank to give a constant head and adjustable restrictors. The sensitivity is calibrated by a subsidiary experiment with a micrometer placed up against the final jet.

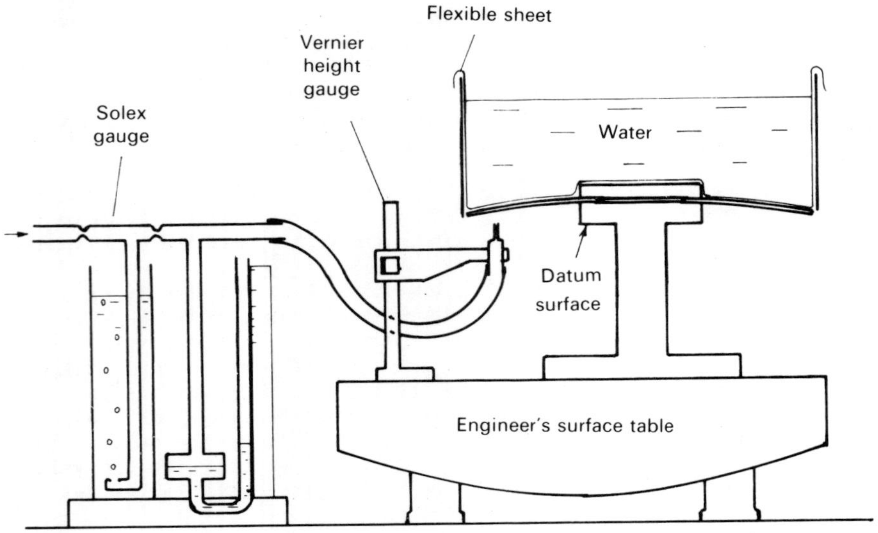

Figure 7·6

The jet may be made from fine tubing obtainable from modellers'
suppliers.

QUESTION

7.1 Figure 7.7 shows the loaded position of a structural model.
To reach this position, the load W was gradually increased to 20 N.
If the friction coefficient between the loading arm and structure
is 0.25

(a) find the horizontal force exerted on the structure

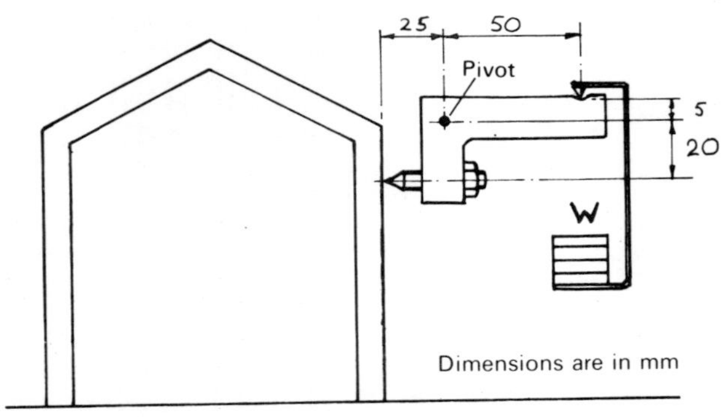

Figure 7·7 (see questions)

(b) find the value to which the load can be reduced before return
     slippage starts

(c) if the structure has a lateral stiffness of 30 N/mm and is
     artificially deflected further to the left by 3 mm, find the
     required load at W to prevent back-slip, if the pivot of the
     loading arm remains position-fixed.

8. GAS ANALYSIS

This chapter and chapter 9 come closer to pure science than the previous ones and serve to illustrate cross-sensitivity effects and the estimation of small errors.

Engineering students will be particularly concerned with combustible gases and combustion exhaust gases. The problems arise not only in analysis of samples but also in the taking of samples.

8.1 SAMPLING METHODS

When a long run of pipe is available, say over 100 pipe diameters after the end of active combustion, and we have a steady process, the gases are likely to be thoroughly mixed, any condensate being in equilibrium with the local atmosphere, that is, saturated with all the constituents at the pipe temperature. In most cases the dry gas only is required; to be sure of a true sample we should avoid getting condensed liquid with it, since on cooling the condensate may dissolve some further gases and distort the composition. Thus the sampling pipe should protrude into the main pipe to avoid collecting run-off from the pipe walls, bearing in mind that the sampling may well be intermittent.

Any small amounts of condensable vapours would eventually clog the sample line; if the sampling line is to remain connected over long periods it is best to have a falling line finishing at a catch-pot and filter chamber.

The inlet is preferably transverse to the flow, not upstream-facing, so that it does not fill up with impacting soot or ash particles. Whether to face it downstream or not must depend on whether we wish to collect such particles during the period when a sample is being taken. These particles will partly lodge in the pipe but some would reach the filter so that they can be examined. If we definitely want to minimise the pick-up of particles, the pipe may as well face downstream.

Pulsed flow as from engines is very liable to be misrepresented by the sampling method. The earliest gas released at high speed when exhaust starts may be of different composition (and *will* be of different temperature) from the later portion. Samples from the exhaust pipe will tend to consist mainly of the latest gas that remains in residence until the next exhaust pulse. For pollution control one procedure is to pass the whole exhaust gas into a flexible bag over a limited period, then let it mix so that a sample represents a true mass-average. The bag can then be sucked empty

before the next test. This is particularly appropriate for tests
that simulate town driving consisting of cycles of acceleration,
over-run, idling and steady cruising in various proportions.

If the gas flow is stratified, it seems most desirable to sample
from several separate points and also make some attempt to measure
the velocities in each layer. Heavy vapours have a slow diffusion
rate, of the order of a few metres per hour in still air, and can
stay segregated for long periods, a serious hazard in boats where
a pool of butane gas can lie in the bottom until a cigarette end is
dropped into it. On a large scale, the same can happen in pipe
trenches and sewers. (Also see section 9.1.)

8.2 OUTLINE OF PHENOMENA AVAILABLE FOR ANALYSIS

The most absolute methods are based on chemically absorbing each
constituent and measuring the volume change. Other chemical methods
are based on colour changes in special reagents or special indicators, electro-chemical changes generating an electrical signal, or
chemical absorption followed by detecting a change in physical properties. The most convenient are methods based on physical properties, the most usual being thermal properties, absorption of specific infrared rays, diffusion rates, magnetic properties, or density.
A limited range of the commoner methods is described below.

8.3 DIFFUSION (GAS CHROMATOGRAPHY)

Although somewhat too specialised for this book, for the sake of
completeness diffusion will be mentioned briefly. The rate of
diffusion of a gas varies inversely as the square root of the molecular weight. For analysis, a sample of gas is·injected rapidly
into a stream of carrier gas such as nitrogen, passing through a
long small-bore tube packed with porous material. The first constituent to arrive at the far end is hydrogen, followed by other
gases in order of molecular weight. This raises a few problems of
overlap (CH_4, H_2O), (C_2H_4, N_2), (C_3H_8, CO_2). These can be overcome
by first burning off the combustibles. The arrival of each batch
is detected by changes of thermal diffusivity, density or flame
ionisation methods; these detectors have to be calibrated to make
them quantitative, using specimen gases with known amounts of the
required constituent and also containing typical amounts of interfering gases that could confuse the result.

8.4 INFRARED ABSORPTION

This method is widely used in engine exhaust analysis. The usual
constituents of engine exhaust gases have some spectral absorption
lines that show very little overlap. The principle used is to
shine infrared rays from a hot filament through a long tube full of
well-filtered, smoke-free sample gas. Each constituent absorbs its
own typical band of wavelength in proportion to the amount present.
To detect how much of any particular gas is present we have to devise a way of ignoring all the other absorptions. The favourite
method is to use a cell filled with the constituent required. This
will respond to the right wavelength, in proportion to the amount

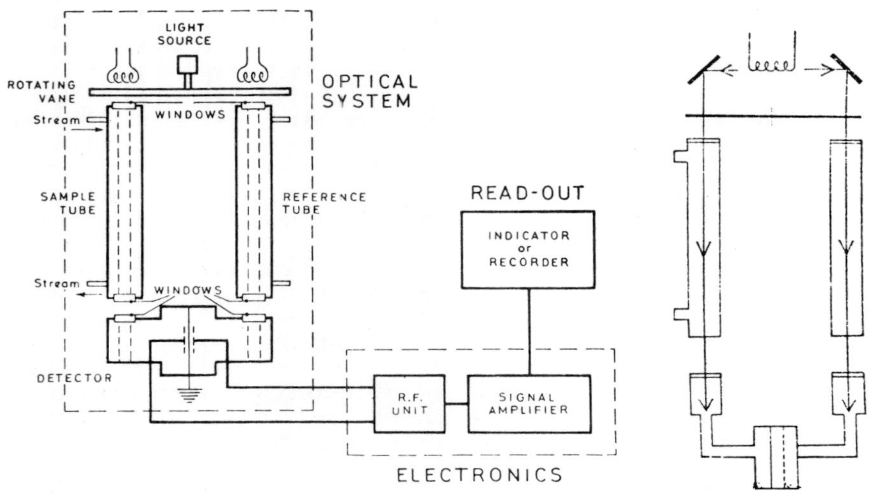

Figure 8·1

not absorbed by the sample. A bridge method is used, one beam
going through the sample, the other going through a reference box
full of plain nitrogen and falling on to a cell as similar as
possible to the measuring cell (figure 8.1). The reference cell
will become warmer than the other because it is receiving a beam
not filtered through the gas sample; in all other respects the
paths are identical. The difference in temperature between the two
cells is used to actuate a diaphragm whose movement is detected
electrically. It is usual to chop the beam in order to obtain a
pulsating signal. This of course makes the signal smaller than a
steady beam but is preferred (a) because of ease of amplifying, (b)
because high temperature rises tend to give non-linear response due
to heat losses. Ideally each gas constituent needs a separate
system, the only common parts being the outer case and the read-out
(indicator, recorder or printer), although one manufacturer supplies
a sub-divided sample chamber, with a purging system to eliminate
interfering gases.

Errors from interaction between the main gases are small but
there could be interference (cross-sensitivity) with other, unusual
gases or vapours. Changes could occur in the sources, mirrors or
windows.

It is normal to recalibrate these instruments at frequent intervals using cylinders of known gas mixtures. Cross-sensitivity may
be tested by deliberately injecting a potentially interfering gas.

8.5 CATHAROMETERS

Despite the advent of the types of instrument described above, there
is still a place for simple flue-gas analysers based on thermal pro-

perties. CO_2 has a lower thermal conductivity than nitrogen or
oxygen. A catharometer cell contains two similar filaments forming
two arms of a bridge; one filament is in air or nitrogen, the other
in sample gas. If both carry the same current, the one in air will
run cooler than the one in gas containing CO_2. The bridge can be
adjusted to give zero output when both cells contain the same gas.
The filament should not be directly in a stream of flowing gas, this
would make it respond to speed of sample flow; it should cool by
natural convection. Both filaments should be at the same pressure.

The main use of catharometers is in CO_2 measurement for combustion control. It is possible to measure CO also by this method,
indirectly. CO as such has properties similar to air; the CO_2/CO
meter takes a sample, passes it through a CO_2 meter, then through
a tube full of copper oxide in a furnace where the CO is oxidised
to CO_2 and finally through a second CO_2 meter. The oxidising tube
has to be regenerated from time to time by passing air through it
while it is hot.

The main trouble with the CO_2 meters of this type is a very
strong cross-sensitivity to hydrogen. A boiler plant is likely to
have a little hydrogen in the flue gas, especially if the air-
supply is insufficient in quantity or not properly directed to the
flames. One per cent of hydrogen has the effect of about minus
8 per cent of CO_2; so if your CO_2 meter reads low or negative
sometimes, check for hydrogen. If it *always* reads negative, you
may have some reversed connections.

The catharometer principle can also be used as the detector at
the end of a diffusion apparatus, as mentioned in section 8.3.

8.6 MAGNETIC INSTRUMENTS

A method of controlling boiler combustion is to maintain a small
excess of air in the flue gas, enough to aid complete burning yet
not enough to cause excessive heat loss up the chimney. We measure
the unburnt oxygen by using its paramagnetic property.

One way of using the effect is to pass a stream of the flue gas
through a catharometer-type cell with two filaments placed symmetri-
cally, with a permanent magnet outside the cell, closer to one filament. When the gas contains oxygen, the magnetic field attracts
additional flow over one filament so that it is cooled preferentially.
The magnet can be arranged to swing out of the way for instant zero
checks. A check of scale value is needed at intervals, using a known
gas mixture, in case the magnet loses strength or the filaments become oxidised or corroded.

An alternative form of detection is to suspend a diamagnetic rod
in the field of a pair of permanent magnets, on a delicate fibre
suspension. The field is arranged unsymmetrically to give a repul-
sion torque on the rod; if paramagnetic gas is admitted to the box
the magnetic field is increased and the rod deflects. The deflection is read by light-beam and mirror.

The magnetic method is cross-sensitive to nitric oxide which is also paramagnetic, with a susceptibility about two-fifths of oxygen. [14] The response can be quite quick; 95 per cent response in a few seconds is feasible with a small mirror-deflection instrument. Thermal instruments tend to be slower.

8.7 ELECTRO-CHEMICAL CELLS

Before proceeding to the section on more traditional chemical methods, it should be mentioned that there are various electro-chemically active systems that take up a reversible equilibrium with the gas sample and can be made to give electrical signals proportional to the amount of the particular gas they are intended for. One of the problems is cross-sensitivity to hydrogen. The electrodes collecting the electrical output are of platinum, known for its affinity for hydrogen. Possibly gold (which is often cheaper) might be less affected.

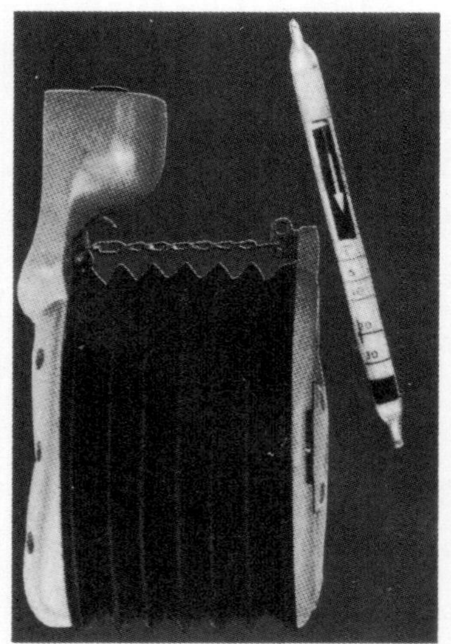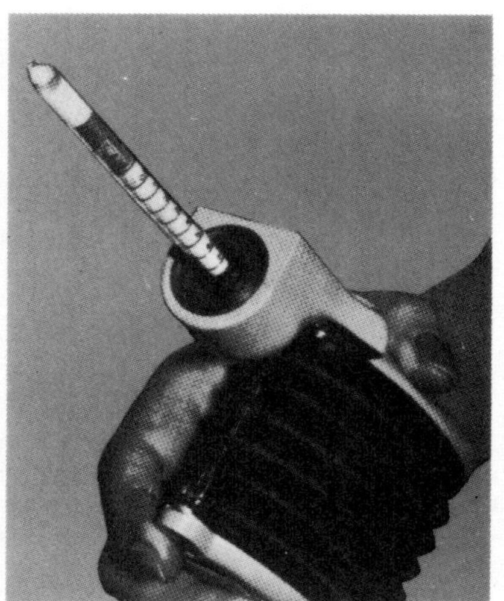

Figure 8·2

8.8 COLOUR INDICATORS

A method of measuring small proportions of gases in air requiring minimal apparatus uses glass phials pre-packed with specific absorbents and colour indicators. These are kept sealed until required, then the ends are broken off and the phial is fitted into a suction pump drawing a known volume of air into the tube. The constituent being sought causes a colour change (black, blue, brown, etc.) of the pack; the extent of the discoloured zone indicates the amount of gas present. A calibration is marked on each tube (figure 8.2).

Balloon diaphragms to
protect solutions from air
Silica tube
with CuO
Dust
Filter
Eye-level line
for volume
readings
Burette
Reservoir
Glass tubes for surface contact
Absorber
Standard 3-gas Orsat
Extended version

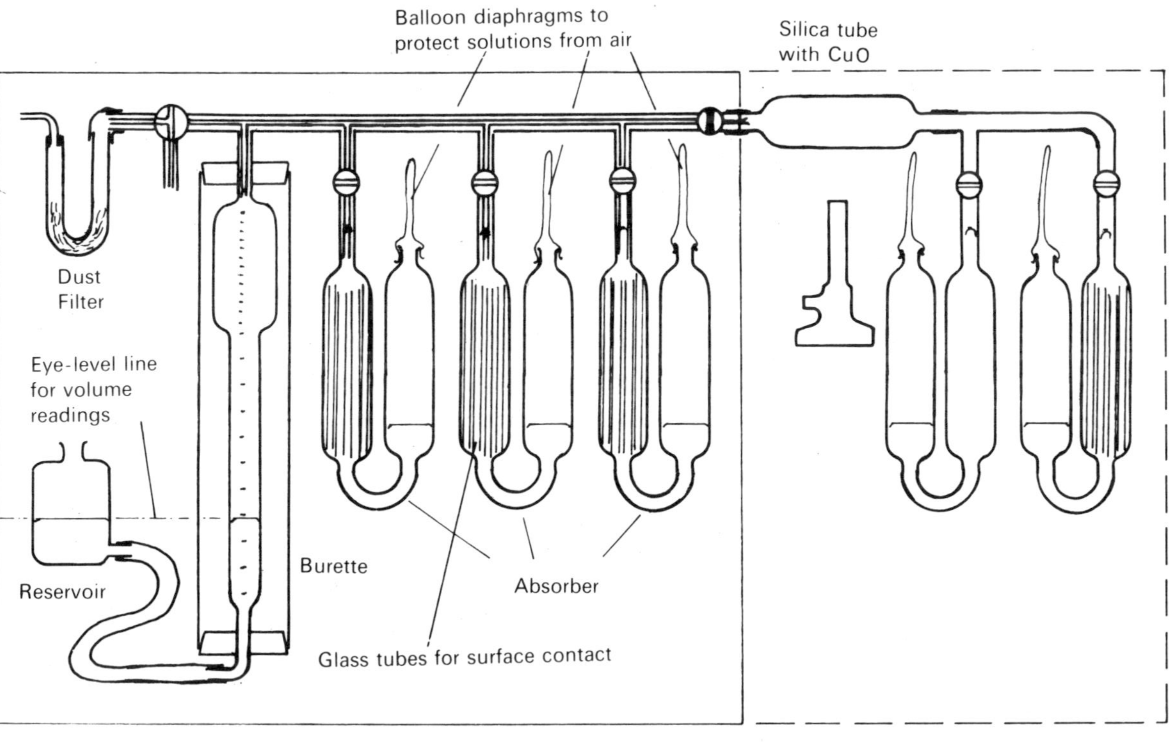

Figure 8·3

They are particularly valuable for safety purposes, detecting harmful proportions of CO and other dangerous gases very clearly, so
that instant checks can be made before personnel are allowed into
doubtful areas such as wells, tanks, unventilated mine workings,
etc. If the ends are sealed off after use, the reading remains
visible for a long time as a record. The makers thoughtfully provide a region of etched glass on which the date and place can be
written. The only practical problem the author has met with these
is when the glass phials do not fit firmly into the pump inlet. [15]

8.9 CHEMICAL ABSORPTION;   ORSAT APPARATUS

The most fundamental method of analysing gas mixtures chemically
absorbs one constituent at a time and measures the ensuing reduction
of volume, at standard temperature and pressure. In the chemical
laboratory there are various forms of such apparatus, differing in
the arrangements for measuring the volume and for performing the
absorption. The engineering student will generally meet the port-
able Orsat apparatus for CO_2, O_2 and usually also CO. An extended
form also measures hydrogen and methane. Figure 8.3 shows both
forms together, to save space. The main parts are a measuring
burette usually with an extended scale and correspondingly reduced
diameter for the lower 40 per cent, separate absorber vessels for
each operation and a moveable reservoir bottle. The burette is
surrounded by a water jacket to reduce temperature fluctuations.
A static jacket is usually enough;   for a bench unit in constant
use, a through-flow can easily be provided.

   The author has considerable personal experience with these devices and recommends the following procedure;   the novice should
take these step by step;   soon the procedure becomes quite easy.

(1) Turn tap to connect burette to atmosphere.
(2) *Gently* raise bottle until burette is full of pink pumping
     liquid. The liquid has considerable inertia and overshoots
     easily.
(3) Turn tap to cut off atmosphere and connect sample line. Draw
     in a sample by lowering bottle. This sample need be only
     roughly a burette-full.
(4) Turn tap back to atmosphere and reject the sample.
(5) Take and reject another sample. This is a quick way of
     purging the sample line of old gas.
(5a) If the sample line is large, turn tap to the position shown
     in figure 8.3. This connects the sample line direct to atmos-
     phere and the line may be purged by self-pressure or by a
     suction bulb. This is also the analysing position with the
     sample isolated in the apparatus.
(6) Now take in a sample to keep. By this time you will be quite
     familiar with the way to turn the tap.
(7) Measure sample volume at atmospheric pressure (see figure 8.3
     for the idea of levelling the liquid surfaces). The sample
     need not be the full volume, this merely saves arithmetic.
     If the volume is off the scale, put on slight excess pressure
     by raising the bottle, open the tap gently *to sample line* and
     let some gas out. Any other procedure is not advisable for

beginners because it is very easy to make mistakes with three-
way taps and draw in some atmospheric air. When sampling hot
gas, wait 2 to 3 minutes and re-measure sample.

(8) Now disconnect the sample line, close it off and take the
apparatus to a comfortable working position.

(9) Open tap on first absorber (clear liquid), raise bottle to
push most of the sample into the absorber. After a few
seconds, draw the sample back. This is called one absorption
and does not get you very far. Draw the gas back into the
absorber which has now become re-wetted on the contact surfaces, bring back again. The author advises six absorptions
in quick succession before attempting a measurement. This
stirs up both the liquid and the gas which helps to get complete absorptions.

(10) Lower the bottle (keep it upright!) until the absorber liquid
is back at the zero mark. Close tap. Measure volume at atmospheric pressure.

(11) You cannot be sure whether absorption was complete, so perform
three more absorptions and measure again.

(12) If necessary, try three more (batches of less than three are a
waste of time). If the reading is still not steady, the solution is exhausted or the apparatus leaks (see below).

(13) Having truly completed the CO_2 absorption, absorb O_2 in the
brown liquid. If large amounts of O_2 are present, this may
take longer than the CO_2, otherwise about the same. Note the
brown film forming on the tube surfaces while oxygen is still
being absorbed. This is a guide to progress.

(14) When truly finished, proceed to the blue or green liquid for
CO.

The chemistry of the solutions makes it essential to finish each
stage thoroughly since any unabsorbed CO_2 will be absorbed in the
next and masquerade as O_2, and so on for O_2 appearing as CO.

There are a few possibilities of large errors; once these are
avoided the method is highly reliable. The first error is absorption in the pink pumping liquid which is dilute H_2SO_4 (about 2 per
cent) and ideally should be saturated with typical sample gas. If
samples are constantly changing to a serious extent, mercury must
be used instead of acid. The second error applies chiefly to the
CO reading. The reaction in the CO absorber has a partly reversible equilibrium. Supposing that several samples of gas with high
CO content have been tested followed by one with low CO content,
the solution tends to give back some CO, producing a falsely low
reading. The author has experienced cases where the give-back was
enough to produce a *negative* reading, a volume increase. This type
of error may be encouraged by changing temperature, so keep the
apparatus away from hot flues or hot sunshine.

The third error is due to gas trapped in the flexible pipe coming
through into the burette between measurements. This is due to lack
of skill and is particularly liable to happen if the sample pipe is
appreciably above atmospheric pressure. For novices it is important
to use transparent piping.

The fourth error can come from trying to save money by making one's own caustic solution. If you wish to do this, remember that caustic soda or caustic potash *pellets* are not just NaOH or KOH but can contain between 20 and 40 per cent water. The recommended solution is usually stated to be 40 per cent W/W of caustic; 20-25 per cent caustic is very sluggish in action.

The fifth error is common with plastics tubing: it becomes slack in time. This can be avoided by putting short lengths of thick rubber tubing over the plastics tubing at the joints to act as garter springs. Rubber tubing rots rather quickly. When changing tubes, cut them off, do not pull them off, clean off residues with lighter fuel.

Leaks may be tested for as follows.

(a) When the apparatus is full, put the reservoir bottle on top and see whether any levels change.
(b) Reject the sample, up to the upper zero mark. Close the tap, place the bottle on the table. Any inward leakage under suction will be visible.

The general causes of leaks are rotten or loose tubes (see above) or lack of tap grease. Too much grease blocks the holes in the taps. Some chemists make up a tenacious film-forming grease by cooking scraps of soft rubber tubing in vaseline.

The calculations are so obvious as to hardly need stating. Note each volume reading. The reduction of volume after each stage is the volume of that particular constituent absorbed; divide this by the original sample volume.

The extended Orsat is for more advanced work with fuel gases, gas producers and analysis of incomplete combustion. It is usual to bring the sample to the apparatus in a pipette as in figure 8.4. The extension comprises a silica tube filled with copper oxide and two more absorbers. The first merely looks like an absorber but is filled with non-absorbing acid, duly saturated with typical gas when high accuracy is required. The second is an additional caustic absorber.

The silica tube holds a considerable gas volume and it is important to ensure that it is full of nitrogen to start with. This may be done by starting with air and using the oxygen absorber several times. It is also important to ensure that this large volume is always at the same temperature when measurements are taken. Some operators do not seem to realise this. The copper oxide can be recharged by passing air over it while heating.

The working of the extended Orsat follows after a normal analysis; the residual gas may contain H_2, CH_4 or other gases. If it is not known what hydrocarbons are present, CH_4 only is assumed.

(1) Open the connecting tap and measure the sample volume; this may differ slightly from the last reading due to expansion or

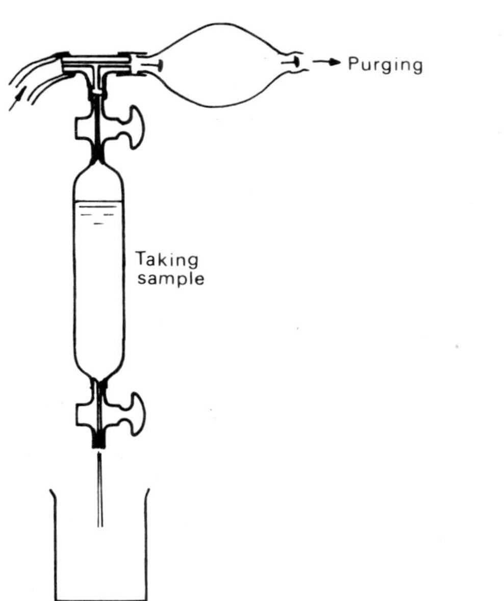

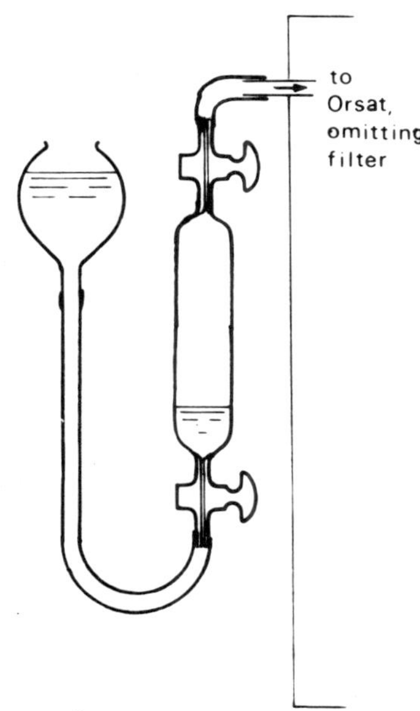

Figure 8.4

contraction of the gas in the silica tube since the previous
analysis. Never mind, take the new volume as starting point.

(2) Open the tap on the acid-filled absorber and heat up the silica
tube to red heat.

(3) Push the gas to and fro six to eight times. Note the volume
reading at an arbitrary pressure, for example, with the bottle
standing on the table.

(4) Push through three to four more times, read again; repeat if
necessary. This is an easy way of detecting completion of the
process of oxidising the combustibles, provided that the silica
tube is at a steady temperature.

(5) When complete, turn off heat and cool the tube with water from
a room-temperature reservoir, meanwhile pumping the gas to and
fro a few times to assist cooling. The cooling process may be
taken as complete when a new steady volume reading has been obtained with the bottle resting on the table.

(6) Now, by lowering the bottle, draw the acid in the absorber up
to the starting mark, close the tap and take a volume reading.

(7) Next, absorb in the second caustic reservoir as in the basic
analysis. Six plus three should prove to be enough to obtain
a repeated reading.

The theory below shows how to interpret the readings. H_2O con-
denses in the reagents.
$1CH_4$ gas + $4O$ from solid CuO = $1CO_2$ gas + $2H_2O$; no volume change
$1H_2$ gas + O from solid CuO = $1H_2O$; one volume lost.

Assuming that the gas contains some of each, the volume change
after the first operation is due to the hydrogen content. The
caustic absorption then gives the methane content since each volume
of methane produces an equal volume of CO_2, by Avogadro's hypothesis.

Suppose now that the gas contains C_2H_4 or C_2H_6.
$1C_2H_4 + 6O = 2CO_2 + 2H_2O$;   one volume *increase*
$1C_2H_6 + 7O = 2CO_2 + 3H_2O$;   one volume *increase*
Thus the method is unable to distinguish between C_2H_4 and C_2H_6 but
indicates their presence;   incidentally, $1C_2H_4 + 1H_2$ or $1C_2H_6 + 1H_2$
give the same result as $2CH_4$ and zero H_2.

9. COMBUSTION-TYPE INSTRUMENTS

The two classes considered here are those used mainly for safety
purposes to detect the presence of combustible gas/air mixtures and
those that burn the sample with air or oxygen to determine the
energy release with some accuracy.

9.1 'EXPLOSIMETERS'

These instruments draw a small sample of atmosphere over a heated
filament, flame traps being provided on the way in and out. The
filament is hot enough to ignite the mixture and if combustion
takes place the filament gets hotter. The rise of temperature is
detected electrically. Such an instrument responds most strongly
to those gases that burn nearest the filament, such as hydrogen.
If they are to give more equal detection of various gases they need
to be sensitive to long flames as well as short ones. They are very
convenient for safety purposes, but attention is drawn to sampling
problems (chapter 8). The danger in coalmines tends to be near the
roof where methane accumulates, while in oil refineries it is near
the floor. In tankers there may be danger high and low even if the
middle reads clear.

9.2 GAS CALORIMETERS

A well-known industrial instrument uses a small burner fed by a jet
of gas from a regulated pressure, drawing its own air supply like a
Bunsen burner and producing thermal expansion of a long chimney.
The chimney temperature will depend on the energy of the gas and
the air/gas ratio. This is constant in the short term owing to the
self-regulating action of the burner but it should be realised that
the airflow depends not only on the driving force available but also
on the resistance to flow which will increase with time as various
surfaces get covered in dust deposits. Dust particles have a pro-
nounced tendency to adhere to surfaces by thermophoresis. Also,
since the natural draught forces are very small, air currents have
a strong effect; up- or down-draughts will change the flow of air
and alter the temperature of the chimney. It is recommended that
these meters be placed in a ventilated cupboard.

The flow rate is regulated by a delicate governor often supple-
mented by a device to compensate for changes of atmospheric pressure
so that the mass-flow of gas remains constant.

Humidity of the air will have a slight effect since the specific
heat of water vapour is 20 per cent greater than that of air and
the air may well range from 0.5 to 1.5 per cent water vapour.

These instruments are used with great success provided they are calibrated occasionally with a known gas sample. The reason why great accuracy is expected of such meters is because public gas supplies are metered by volume but paid for in terms of energy (therms); a small error can mean large amounts wrongly charged.

A more fundamental instrument is the Boys'-type calorimeter in which the heat released by the gas is measured by observing the temperature rise of a measured amount of water, used in conjunction with a positive-displacement gas meter. This device has a Bunsen burner which induces its own air-supply, not metered, but it has a long heat-transfer path in counterflow with water initially just slightly below atmospheric temperature. The operator should regulate water and gas rates so that the emerging burnt gas and excess air come out at room temperature. Then it does not matter what the airflow rate is. If this is not achieved, the heat carried away is relatively difficult to find. The airflow rate is likely to vary with age, decreasing gradually as the passages become coated with deposits. Some estimate can be made by measuring the CO_2 content of the emerging gas, but the typical Boys' calorimeter has no definite outlet flue - the gases emerge from a number of separate points.

When used with the correct flow rates to give atmospheric outlet temperatures, the only unknown error is the (small) heat loss from the casing to atmosphere which is not readily measurable. The rest of the equipment should be as accurate as the mass-flow measurement of gas and water and the water temperature measurement. As noted in earlier chapters, the water should be weighed rather than measured, with precautions against evaporation. The water temperature measurement should be largely free from thermal contact error because water is an excellent heat transfer medium and has direct contact with the thermometer. Thermometer calibration error could be estimated by interchanging the ingoing- and outgoing-flow thermometers. However, the instrument may not be designed for this elementary check to be feasible during the experiment so the thermometers must be compared in a common bath at both temperatures.

The volume of gas is in doubt because of the moisture content. Coal gas is relatively moist at the source and deposits some of its moisture in the pipes underground; it will thus tend to be saturated at earth temperature which will vary with place and season. Petroleum gases straight from a high-pressure main or from dry storage will be much drier. The wet meter will moisten the gas partly, leaving room for an uncertainty of perhaps 0.5 per cent. It seems wise to humidify the gas fully before it reaches the meter.

9.3 LIQUID AND SOLID FUELS

The regular method for these is to burn them at constant volume in compressed oxygen, in a vessel pessimistically called a bomb. The bomb is usually of bronze; copper-based alloys are safer with oxygen than steel, especially if there is risk of hot spots due to friction, impact, oil splashes, etc.

The procedure is to place a carefully weighed amount of the fuel

in a crucible in the bomb, with provision for electrical ignition.
The bomb is quite small for several reasons and is filled with
oxygen at considerable pressure.

It is immersed in a water bath fitted with a power-driven
stirrer. The temperature of the water is noted before the fuel is
ignited and then monitored until the end of the experiment.

There are some variations in the way of taking care of heat
losses. The first priority is to minimise these.

The heat loss by evaporation is kept small by closely covering
the apparatus with a substantial lid of low thermal conductivity
and preferably of low heat capacity. It is traditional to use wood;
a modern calorimeter could logically use a low density closed-cell
foam.

The loss from the sides is minimised by making the water bath of
highly polished copper to cut down radiation (do not lacquer the
copper, lacquers have high emissivity in the infrared; you may however gold-plate the bath if you wish).

In addition to this, the bath is surrounded by a guard-bath kept
at a temperature close to that of the calorimeter bath. A small
air gap is maintained between the baths with the help of small
spacers of cork.

Here there are several possible procedures.

(a) Keep the outer bath temperature under observation; keeping it
    uniform by stirring, keep a record of the difference across the
    gap and find the heat flow at a known difference (see below).
(b) Vary the temperature of the guard-bath in step with the calori-
    meter bath temperature. This reduces the heat-flow but stops
    it from being estimated. Obviously this is not a good method.
(c) Use procedure (b) but keep a continuous record of the difference
    between the baths.

To measure the errors, observe the temperature for some time before igniting the fuel, keeping the power stirrers in action. Then
after the combustion is finished, if using procedure (a) it is
merely necessary to keep reading both temperatures for some time.
If using method (c), allow a certain typical temperature difference
to arise between the baths. Keep recording both temperatures with
stirrers kept going.

The use of this procedure is fairly obvious. If the apparatus
is at room temperature at some instant, the rate of rise of inner
temperature before ignition gives the stirring power entering into
the system, minus evaporation loss.

After ignition, the fall in temperature indicates the net heat
loss from the calorimeter at the particular temperature difference.
If there were no heat loss, the temperature would have continued to
rise at the original rate due to the stirring.

In method (a), a fair correction can be made directly from the time-temperature graph if the lines are straight enough and if the temperature rise is fast enough, by extrapolating both curves to one instant of time (figure 9.1). This does *not* measure the lid error, which is mainly a thermal capacity effect.

Temperature

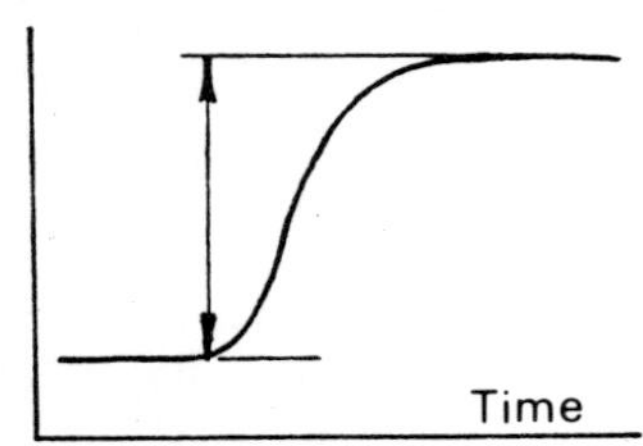

Figure 9·1

The other errors in this method are quite small, being the heating value of the igniter wire, the electrical igniter input, products of combustion at the final temperature, uncertainties of heat capacity (water equivalent) of the metal parts, lack of care in weighing the calorimeter water, ensuring that it is all poured into the bath, etc.

By due precautions, everything except the lid error can be brought under control to such an extent that it seems wise to arrange for duplicating the thermometer in the inner bath, thus obtaining some check on uniformity of temperature.

If all the other variables have been measured, it would be possible to estimate the effect of lid loss by repeating the experiment with the same fuel but pre-heating the lid before use, thus reversing this particular heat flow; the pre-heat should be more than the final bath temperature because of the loss of time while fitting up the equipment.

The answer includes the chemical energy from oxidation of igniter wire and any cotton wick used but excludes latent heat of water vapour, since the final gases are likely to be saturated with water vapour, the greater part of the water having condensed. The correction can be calculated from bomb volume and steam tables. If excess fuel was used the reaction will be incomplete. The result will *not* be the calorific value of the fuel but that of the oxygen.

10. CASE STUDIES AND OTHER EXAMPLES

These remarks are intended to be helpful to students preparing to enter into research work but should be of some interest to all students.

10.1 TWO CASES OF MISSING SUBSTANCES

In the first case (figure 10.1) exhaust fumes from a combustion process were being measured. It was important to account for the ingoing substances which involved measuring and analysing what came out, the most significant uncertainty being water content. It was

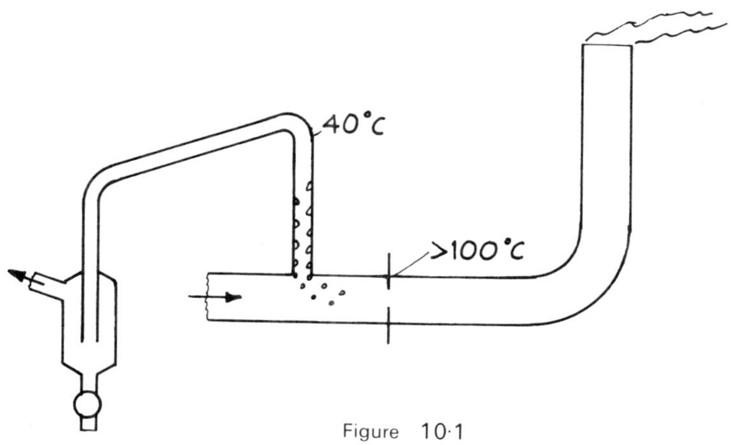

Figure 10·1

evident from the clouds of white smoke that the amount of water was substantial. The sample pipe rose vertically upwards from the horizontal main pipe to about 1 m above, then fell to a vessel in which the moisture collected, to be drawn off and measured at intervals. The dry gas sample flow was filtered and measured also, giving the water/dry gas ratio. Curiously, this was extremely low. The reason was that the highest point in the sampling line was at ∿40 °C, so that the vast majority of the water condensed in the rising leg of the sample line, dripped back down the inside wall and was blown away with the main flow. Due to loose fittings some water also dripped down the outside.

This error was discovered too late for action to be taken; the data had been published and it would have been too confusing to re-

pudiate them. The ultimate result was that the importance of excess
moisture in the reaction was never fully recognised.

The second example arose during a combustion experiment concerned
with smoke emission. A small chamber was burning raw coal, the
smoke density being observed by light-beam and photocell (figure
10.2). On one occasion the photocell registered negligible smoke
but looking at the chimney large amounts of dark smoke were seen.
The reason for the low smoke reading was that at the instrument the
smoke was at over 400 °C, all the tarry matter being in the vapour
state and therefore transparent.

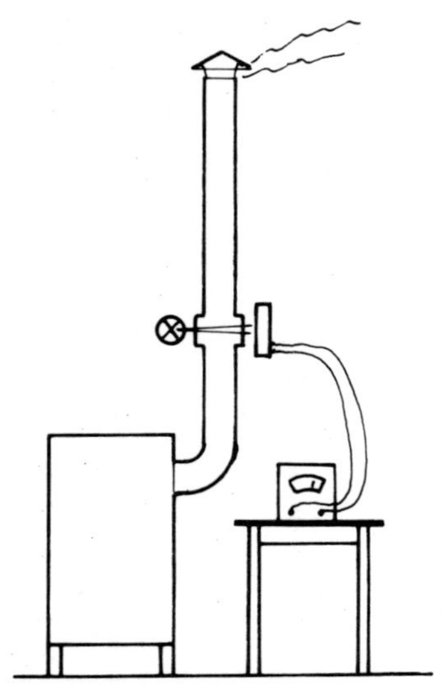

Figure 10·2

It was fortunate that this experiment was set up in temporary
premises, with its own chimney so that the final result was access-
ible. In a respectable well-equipped research building the smoke
would have passed into a common stack and been mixed with other
effluents; the evidence would have been obscured.

10.2 CONVENIENCE BEFORE RELEVANCE

When burning raw coal (and various other substances) it is well established that smoke-free burning, producing CO_2 and H_2O and very
little else, can be achieved by ensuring that all the carbon particles and volatile (tarry) matter are made hot enough and given
sufficient air. This is not easy to achieve at all times and an
investigation was made to see whether a mixture of cool smoke and
air in suitable proportions could be made inflammable, thus burning
off at least some of the noxious products. For systematic experi-

92

ments a constant, repeatable and controllable source of smoke was
needed. A convenient method was found in the partial pyrolysis of
methane, passed through a red-hot tube. The resulting smoke when
mixed with air was found to be readily ignitable without pre-heat.
What was not apparent was whether the ignitability was due to the
mixture of air and smoke or due to the hydrogen and residual methane.

10.3 UNREALISTIC APPARATUS

The example given here comes from the fascinating subject of thrust
augmentation by ejector action. Propulsion by high-speed fluid
jets is inefficient at low forward speeds. A thrust augmentor is a
venturi into which a high-speed jet is injected. This entrains air;
the low pressure at the throat can be taken advantage of to increase
the speed of the forcing jet and the venturi duct is shaped to obtain thrust from the induced airflow.

The effect was used for propulsion in war-time on a limited scale
and many experiments have been conducted. It is relatively easy to
obtain 20 per cent thrust gain. By using a wall-jet (Coanda) or sub-
divided jet (author) the gain can exceed 30 per cent. The Lockheed
Humming-bird aircraft used the system for vertical take-off and
hovering, with 40 per cent gain. Reeves Morrison and others used
multi-stage jets to obtain gains of 70 per cent or more (figure
10.3a, b and c).

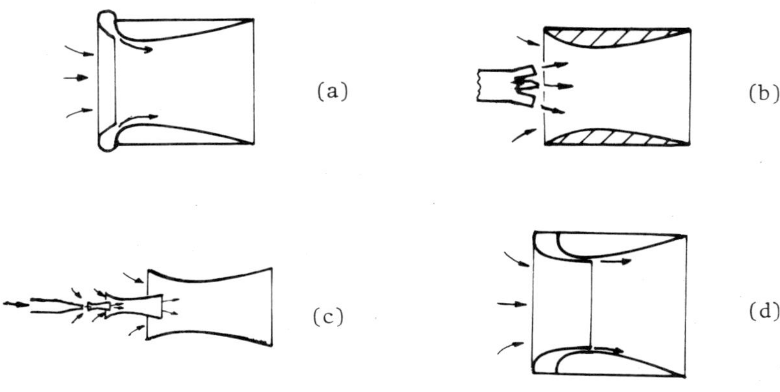

Figure 10·3

The investigation discussed here used a wall-jet system (figure
10.3d) and found a thrust gain of only about 10 per cent, compared
with the direct thrust obtainable from the same forcing air mass-
flow at the same supply pressure but without entraining extra air.

On looking into the details, it was noted that the wall-jet construction gave a very long narrow guide channel, far longer than
needed to give straight flow and involving considerable friction
losses, estimable at over 10 per cent of the available pressure;
thus the thrust gain discounting this loss would be more like the
usual 20 per cent.

10.4 INVESTIGATION VERSUS DESIGN

The example shown here concerns a quick-release aircraft coupling
(figure 10.4). The original design had a reduced central portion,

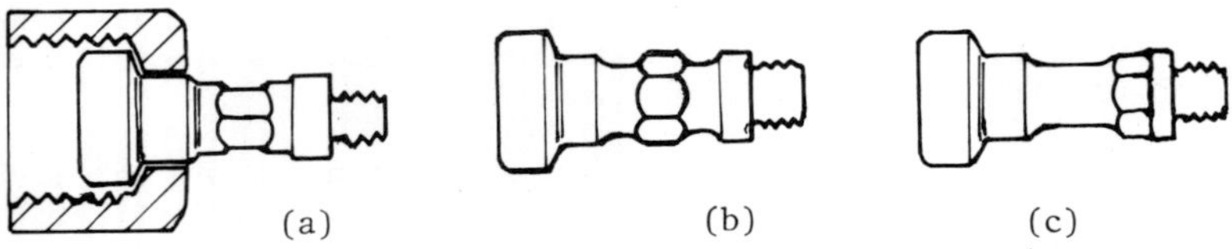

Figure 10·4

presumably to save weight. Spanner flats were provided for assem-
bly purposes, cut into the reduced portion (a). A lengthy programme
of detailed research produced changes giving double the fatigue life.
An orthodox sound design would have used methods (b) or (c) with very
little additional weight, keeping the spanner flats away from the
reduced portion.

10.5 NOT LOOKING FAR ENOUGH

There have been at least two published cases roughly as follows.
Chassis frames for road or rail vehicles are sometimes built on the
Vierendeel truss or ladder-frame principle of flexible bars disposed
at right-angles with occasional stiffeners. A problem with these is
cracking at the stiffeners and there have been successes in reducing.

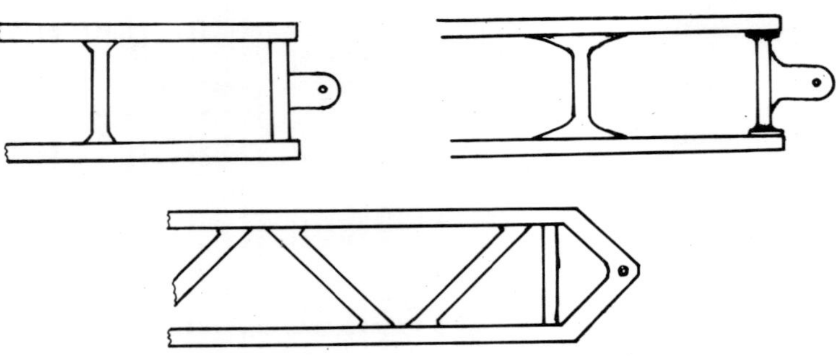

Figure 10·5

94

the cracking by taking metal away rather than adding metal. It is
of course true that in a redundant, flexible structure the stiffest
force-path gets the greatest forces. The remedy of taking metal
away is a well-known repair expedient (drilling the head of a crack).
As a design cure, removal of metal is only valid if the stiffness
was not in fact necessary in the first place. Another approach is
to add stiffness evenly all over or radically alter the structural
concept (figure 10.5).

10.6 A CONFUSING SITUATION

The conversion of fluid velocity into pressure in a diffuser gives
rise to good instructive experiments. It is also the subject of a
great deal of research. The following remarks are intended to show
the student the wider implications (or complications) involved in
applying a simple principle.

The chief uses of diffusers can be broken down into

(a) the diverging part of a venturi, used for flow measurement, for
    atomising (carburetters) or for relative slip (gas-scrubbers).
(b) pipe-work transitions especially after pumps or fans, including
    outlets to free space
(c) aero-engine intakes in high speed flight and at take-off (dif-
    ferent approach conditions)
(d) internal diffusers in turbomachinery.

Many of the variables are pre-determined by the conditions of
use or are economically restricted; they may include many of the
following.

(1)   Over-all length available in terms of inlet size.
(2)   Over-all area ratio.
(3)   Shape (circular, oval, rectangular).
(4)   Approach velocity (sub-, trans- or supersonic).
(5)   Reynolds number on diameter and on length basis, in other
      words flow profile (plug, fully developed, boundary layer
      thickness).
(6)   Swirl, pulsations, non-uniformity especially in (d).
(7)   Direction not in line.
(8)   Pipe axis not concentric.
(9)   Change of size, deliberately or in error.
(10) Protruding flange packings or recesses.
(11) Transition curve from straight to divergent zone: length,
      law (arc, $y = ax^3$, $y = ae^x$,...) in terms of diameter.
(12) Transition curve considered in terms of boundary layer.
(13) Initial divergence angle (determines rate of deceleration
      in conjunction with diameter).
(14) Later divergence angle (convex or concave walls).
(15) Transition shape into next pipe.
(16) Pressure recovery in next pipe.
(17) Any restrictions in next pipe preventing harmful back-flows
      (for example, blade rows, exit filters).
(18) Any air or vapour bubbles to be compressed (cavitation, open
      discharge, pipe not full, ...).

(19) Any abnormalities in fluid, for example, drag-reducing addi-
     tives, dust, ...
(20) Any diabatic effects, for example, heating, cooling, evapora-
     tion, condensation, icing (carburetters?).

This should show how much thought is needed to devise a representative experiment for a given condition. For example, it is already
known that for some common values of items (2) and (11) the best
values of (13) range from 15° to 6° as item (1) ranges from 2 to 20
provided (15) and (16) are inapplicable.

10.7 MISLEADING THE UNCRITICAL

We tend to forget that students are not familiar with the context of
specialised topics. Just one example should suffice to illustrate
how a statement can mislead by being true but insufficient.

In heat exchangers we often use an arrangement of banks of tubes,
with one fluid inside, the other passing transversely over them.
This is technically a cross-flow arrangement although usually the
external fluid weaves to and fro over the same tubes several times
with the help of baffles so that ultimately a good degree or counterflow is obtained. (Sometimes we choose parallel flow at first, when
we wish to minimise tube wall temperatures by putting the hottest
fluid outside near the coldest internal fluid.)

The point at issue here is the detailed layout of pipes (figure
10.6). It is found that *for a given fluid speed* the staggered
arrangement (a) (as expected) gives higher heat transfer rates. We
imply that this extracts more heat or saves on tube material and is
altogether better.

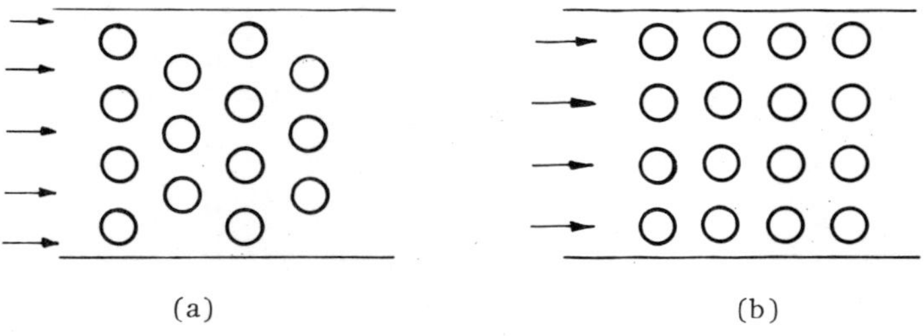

(a)   (b)

Figure 10·6

When, however, we look more deeply we must consider the cost of
producing the given speed. It is known that the pressure loss with
staggered pipes is consistently greater than that with straight rows
(b), in many cases almost double. The heat transfer is in most cases
only 20 per cent greater. Now, if we balance the saving due to less
pipes in the staggered set against the greater pumping pressure, in
plant cost and more particularly in energy cost, it will often be
better to use straight rows.

96

It is interesting to note here that heat exchangers are a good example of optimising; many engineering decisions have a very flat least-cost curve and a small shift of one parameter, for example, probable future energy prices, causes a very large shift of the computed optimum solution. A good example is given by Hawthorne [16] from which figure 10.7 is taken.

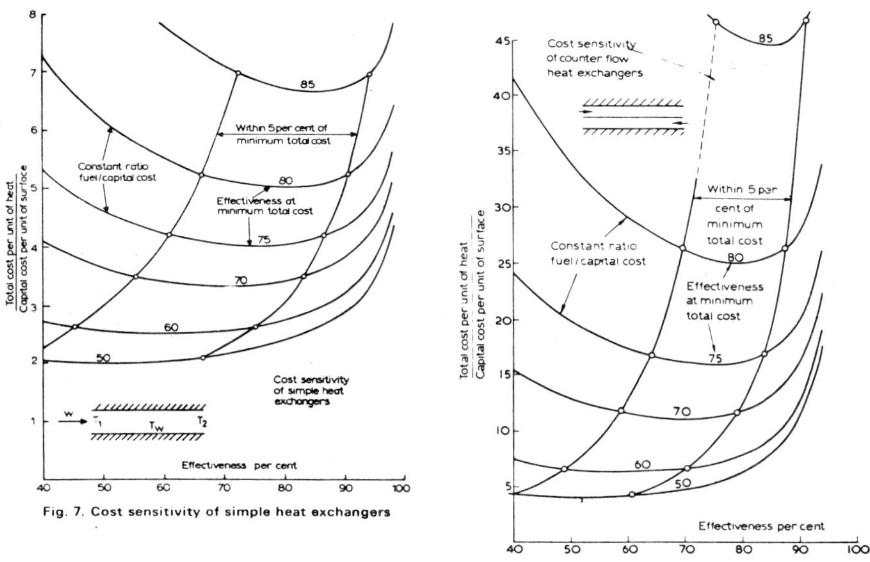

Fig. 7. Cost sensitivity of simple heat exchangers

Fig. 8. Cost sensitivity of counterflow heat exchangers

(Reprinted by permission of the Council of the Institution of Mechanical Engineers from Hawthorne [16].)

Figure 10·7

Fortunately it is becoming more usual to take note of this and present the results of optimising calculations with an indication of relative sensitivity to the more difficult assumptions and with an indication of the relatively insensitive plateau.

APPENDIX I

RESIDUAL STRESSES, PRIOR LOADS

The point at issue is how much residual stress is produced by a
given loading, how it affects service behaviour and how it interferes with fatigue test results. The majority of practical cases
relate to tensile loading and residual compressive stresses. One
set of data comes from Taylor, [7] one of the most significant
papers in the literature on pressure vessels. A summary is given
here of the most directly relevant parts.

Specimen pressure vessels with branch pipes ('nozzles') were designed with suitable reinforcement, to a realistic scale (20 in.
vessel diameter, 7 in. pipe, thicknesses 1 in. and $\frac{3}{8}$ in., $1\frac{5}{8}$ in.
reinforcement). Several of these were built and fatigue tested,
each specimen being tested to a given pressure, repeated until leakage started. Some of the specimens were taken above the design
value, giving lives < 40 000 cycles, others to considerably lower
pressures (figure A1.1).

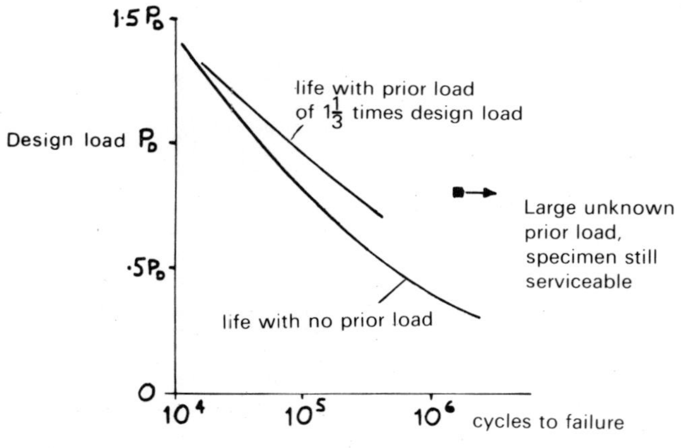

Figure A1·1

Some further similar specimens were given a prior pressurising
to $1\frac{1}{3}$ times the design pressure (this is the acceptance test pressure in some codes, though $1\frac{1}{2}$ times is more traditional), followed
by fatigue tests to various pressures.

One vessel, by a lucky accident, was tested to a greater pressure.
Though this pressure was not recorded, the fact that the vessel sur-

vived this pressurising suggests that the value was not more than $2\frac{1}{4}$ times the design pressure.

The graph shows the basic life curve, the curve for vessels prior-loaded to $1\frac{1}{3}$ times design value which survived slightly longer, and the odd vessel which survived longer than any of the others at comparable pressures. At 2 million cycles it was well set for indefinite endurance barring corrosion-fatigue.

Another piece of evidence is due to Heywood; [17] this covers a wider range of prior loads. Tensile specimens with a small transverse hole, stress concentration factor ∿3, were fatigue tested over the same range of stress but each specimen was given a different prior load, some tensile and some compressive. The original presentation is somewhat confusing to the beginner; figure A1.2 presents

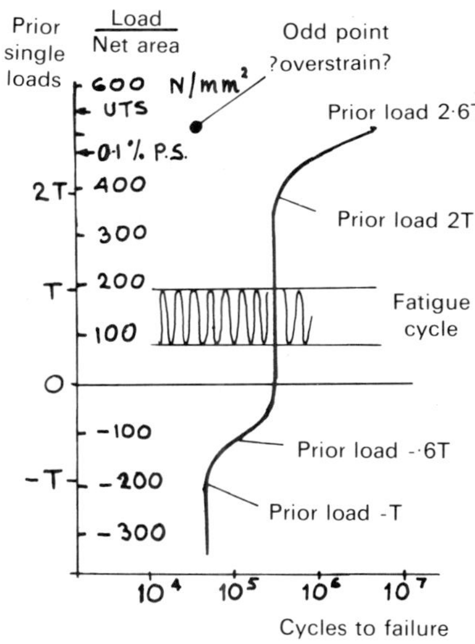

Figure A1·2

it in a slightly more explicit and generalised form. In this case the tension ranged from nominal stresses of 192 to 83 N/mm². The material was an aluminium alloy of 0.1 per cent proof stress = 465 N/mm². The actual local stresses cannot be spoken of without considering the consequences of yielding during the prior loads. Some specimens were pre-loaded to 500 N/mm² nominal, others to lesser tensile or compressive values (see graph). This is perhaps easier to follow than the Taylor case since the service load is the same in all cases, only the pre-treatment being different.

It is seen that prior loads become very favourable when over 2
times the service load, by which time considerable local yielding
must have occurred. In reverse, harmful life-reducing effects are
noted.

To attempt a full analysis would be too lengthy for the present
purpose but an illustrative example is given (figure A1.3). Assuming an ideal elasto-plastic material of yield point Y, let us make
a component with a stress concentration factor of $2\frac{1}{4}$, in repeated
loading from zero to a general stress of 0.4Y. These are typical
of pressure vessel details designed for fatigue loading.

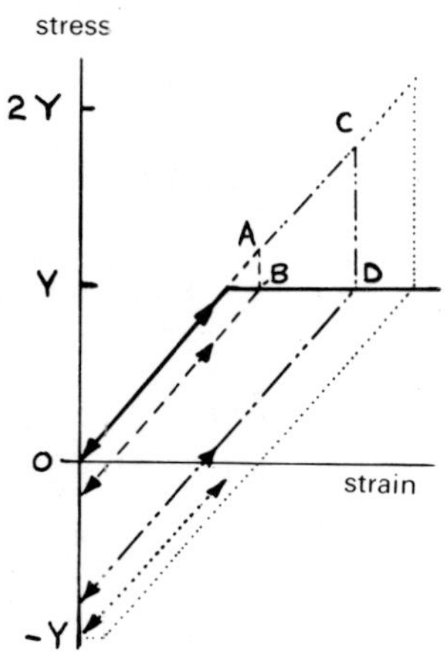

Figure   A1·3

Without prior loading the situation will remain entirely elastic,
the local stress cycling from 0 to 0.9Y. This may be marginally
safe for fatigue provided the welds, etc., are fully stress-
relieved, depending on the material concerned.

A prior load of $1\frac{1}{3}$ times design value will put the general
stress up to 0.533Y; the local stress tends towards point A (figure
A1.3) but cannot get there, remaining at stress level Y; the amount
of yield will correspond to the missing length AB, namely 0.2Y,
(since 0.9Y $\times$ $1\frac{1}{3}$ = 1.2Y).

When we release the load, the main body of the component will
force the local stress into compression by $\sim$0.2Y. In the subsequent
fatigue cycle of 0 to 0.4Y the local stress will have a range of
$2\frac{1}{4}$ $\times$ 0.4Y, thus its limits are -0.2Y to +0.7Y. This is manifestly
slightly better than the first case.

Now consider a prior load to 2 times design value. The elastic
stress at the stress raiser tends towards point C at 1.8Y. The
missing stress corresponding to the yield will be CD = 0.8Y; this
now will turn up as a residual (compressive) stress of -0.8Y.

The working cycle has now shifted down to the levels -0.8Y to
+0.1Y, with a substantial compressive mean stress; clearly this is
much more favourable for long fatigue life.

An even larger prior load would not achieve very much more, as
indicated on the diagram. The residual compressive stress will be
$\sim$-Y.

APPENDIX 2

FURTHER THOUGHTS ON FRACTURE TESTING

A2.1 STRESSES

Considering the three specimen types shown in figure 4.4, an attempt
has been made to show the variation of loads needed to give the same
maximum stress on specimens of equal net areas.

There are several approaches to calculating the stresses. The
first specimen may be considered to have uniform tensile stress
times a local effect $\propto \sqrt{}$(notch depth $\div$ tip radius).

The second specimen may be treated as tension plus simple bending
due to the offset of load points from the centre of the net area.
This is again magnified locally (there are more elaborate solutions
available).

The third specimen is more like two cantilevers plus a curved
beam (tuning fork). Curved beam theory takes note of the higher
strains at the inner edge, where a shorter original length has to
accommodate a given extension.

The stresses obtained thus can differ considerably from the stress
obtained by plain bending of a straight bar times notch effect. For
curved beam theory see, for example, reference 18.

A2.2 STRESS GRADIENT

Having noted the consequence of stress gradient and grain size in
section 4.1.1, it seems difficult to avoid the thought that it
should apply here also, so that there may well be a size effect.
This is generally realised and testing is done on specimens of realistic size. To avoid large forces in the machines the temptation is
to use the deeply notched specimens, despite their larger stress
gradients and some loss of comparability.

A2.3 ENERGY TRANSFER

The question of energy transfer from adjacent parts during crack
advance has already been raised briefly. This of course is widely
realised and the student will find references in the literature to
hard and soft machines, though the point of the energy available
from large, evenly stressed real-life structures is not often discussed.

Figure 4.4 makes the point of a hard or soft machine clearer by
using the same machine on two specimens of equal net area. The

machine is hard towards one specimen, soft towards the other. The
soft machine can give the stiffer specimen more energy. The situation is not confined to the striking case of the two extreme specimen types; to some extent specimen thickness alone has this type
of effect, as illustrated by question 4.3.

The reason for the discrepancy between test results on thick
and on thin specimens is generally said to be due to the difference between plane stress and plane strain conditions. If lateral
strain is inhibited by adjacent material, the stress is increased
approximately in the ratio E' / or $1 / (1 - \nu 2)$, giving an
increase of about 8 per cent with steel (see chapter 3). In more
complex configurations the expression sometimes given is more
complicated but still produces discrepancies of around 10 per cent
whereas the test figures that brought the problem to light were
rather larger. It seems advisable to record actual machine loads
and deflections in order to assess the influence of energy transfer.

How feasible it is to simulate real-life cases is beyond the
scope of these remarks. Clearly there must be some limit set by the
size and dynamics of a structure; we would not expect strain energy
at one end of a suspension bridge to affect crack propagation at the
other end; the effect must be localised to some extent.

M.MALEKZADEH

APPENDIX 3

INERTIA LOADS AND STRESS WAVES
- A SHORT REVIEW

If we apply a given force or load slowly, inertia effects are negligible. If you walk slowly on to a spring-board it deflects progressively. If you run on to it quickly it deflects further and bounces back. If you jump on to it with initial speed, it bounces further still.

The same can be viewed more systematically with a mass-spring system of any kind. If we apply a force slowly, we can regulate matters so that there is no overshoot.

If we want quicker motion, we find that we must apply more force, to overcome the spring and also to accelerate the mass. As our force builds up to its maximum we have put in extra work, the mass has kinetic energy and overshoots.

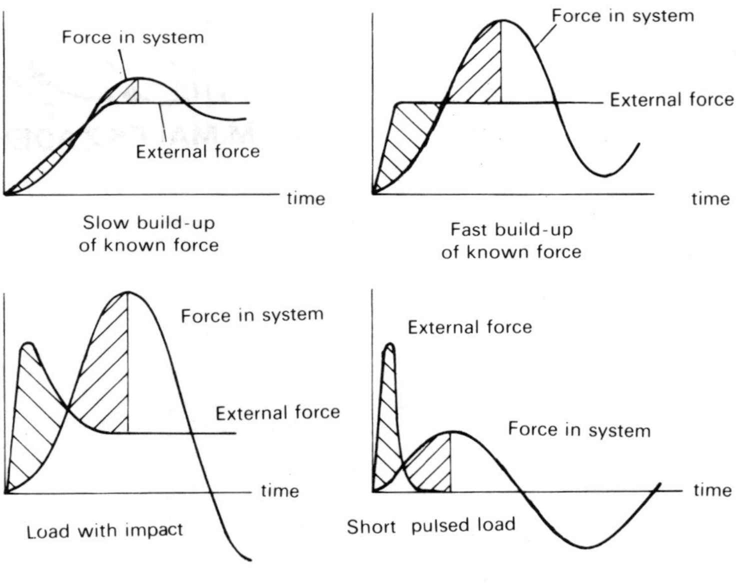

Figure A3·1

Figure A3.1 shows a moderate and a fast build-up of force. The excess force times its period of action amounts to an impulse equal to the area between the curves. This impulse gives the system momentum; it overshoots until the opposite momentum equals it.

This point gives the peak force and peak deflection of the system.
Thereafter the system oscillates about the new equilibrium. We can
only draw these graphs if we know the magnitude of the applied force,
such as your weight on the spring-board, the gas pressure in an
engine, etc.

Another common case is a short pulse load. A car goes over a
bump; the front wheel goes up and compresses the spring for a short
time (so does the rear wheel a little later). The force-time inte-
gral gives upward momentum. The car oscillates but this time about
the original level since the load is unchanged; the input force has
vanished. The magnitude depends on the bump and on hopping of the
wheel.

The wheel and its associated parts, called the unsprung mass for
convenience, also receives upward momentum via the pneumatic tyre.
Its mass is much smaller, the tyre is a stiffer spring. If we had
no dampers the wheel would bounce up high and get pushed back by the
spring. On a bad road the contact time could be less than 20 per
cent of the total, raising problems with steering, traction and
braking.

The forces can be calculated if we can establish the properties
of the contact zone (chiefly stiffness and damping) and the energies involved. The three basic forms of contact are shown in
figure A3.2. The first is general elastic contact. The second can
represent a pre-loaded buffer, a vehicle with a crumple-zone, etc.
The third is a hydraulic or pneumatic buffer with a fixed-size
orifice.

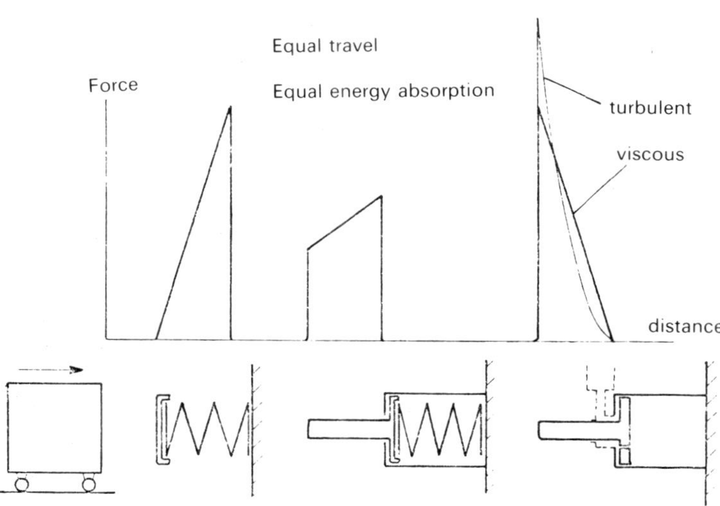

Figure   A3·2

An impact gives rise to stress-waves. The detailed form of the impact determines the form of the wave front or pulse. In the following, a purely elastic impact is considered, with infinitely fast rise for simplicity.

Consider a rod of length ℓ, cross-sectional area a, density ρ, Young's modulus E, fixed at one end. The other end is hit by a rigid hammer of mass M approaching with speed u_o.

The hammer at first only has to accelerate an infinitesimal piece of rod, thus the rod end acquires the velocity u_o. This velocity is passed on as a wave; the speed of compression waves in rods is $C = \sqrt{(E/\rho)}$ and the initial amplitude of the stress wave is $\rho u_o C$. These are all well-established relationships and may be found in many books on elasticity.

The wave travels to the fixed end with undiminished height and is fully reflected. Behind the wave, the hammer travels at diminishing speed. When the wave returns, the next reflection is determined by the new, lower hammer speed.

The stress behind the wave front is diminishing because the hammer is slowing down all the time and pushing less hard. The maximum stress in the bar is not equal in all places; by the time the next wave comes along the previous stress has died down to some extent.

The greatest compressive stress in the rod $\simeq \rho u_o C[1 + \sqrt{(M/m)}]$ (where $m = \rho \ell a$) and comes at the fixed end.

We may expect the number of reflections to be $\sim \pi \sqrt{(M/m)}$ before the hammer bounces away. [19]

APPENDIX 4

HEIGHT ERRORS IN FLOW MEASUREMENT

When measuring fluid flow by pressure drop, what we need to know is
the pressure change due to the velocity; when the two tappings are
not at the same level there will also be a pressure difference due
to gravity head. In many cases the lines to the manometer are filled
with the same fluid as we are measuring; this cancels out the error.
The main source of error is when gas-purged lines are used to measure
liquid flow. A minor error can occur when there is a great height
difference between flowmeter and manometer and the lines are at different temperatures.

The first type of error is illustrated in figure A4.1. The horizontal venturi meter carries liquid, the measurement is by a meter
which must not be in contact with the process liquid for some good
reason, therefore the lines are fed with air, from separate supplies
of course, at a slow rate. Note that this is believed to be undesir-
able anyway because of the meniscus forming when bubbles emerge into

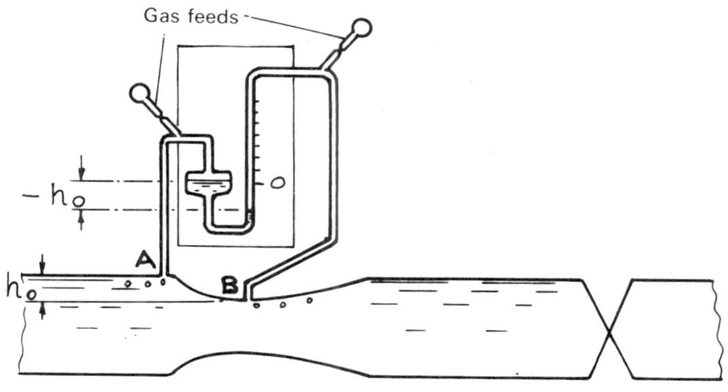

Figure A4·1

high speed flow which creates pulsations. An orifice meter seems
preferable. The static pressure at A is less than at B if there is
no flow simply because A is higher. Thus the meter indicates a
higher pressure at the throat, which is impossible whatever the flow
direction. If the installation were one where there never is zero
flow, as in drainage work, the error could escape notice.

A rather larger error of this kind is shown in figure A4.2a where a vertical orifice plate meter carries liquid flowing downwards. It could equally apply to upward flow but downward flow makes a more striking example.

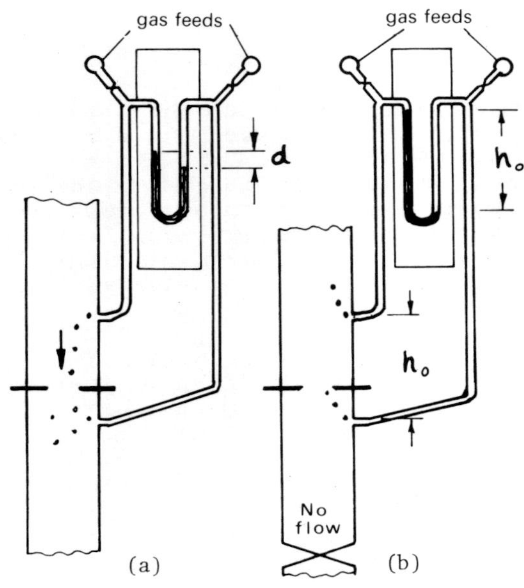

Figure A4·2

Again, gas-purged lines are being used. Suppose at some flow rate the reading is -d (negative), suggesting that there is *upward* flow. If we imagine the flow stopped (figure A4.2b) we see what the true zero should be.

The head relevant to the velocity was $h_o - (-d) = h_o + d$, the actual flow being *downwards* as mentioned above.

Sloping pipes will have the same problem to a reduced extent.

Examples of reasons for using gas-purged lines are as follows. Long lines not easily kept full due to bubbling (beer?). Meter above line, danger of vapour-lock at sub-atmospheric pressure (solvents, petrol). Hot process fluid liable to solidify wholly or partially in the lines or in the meter (sugar solution, crude oil, bitumen, margarine). Fluid carrying suspended solids that would settle out in the lines. Toxic fluid not permitted in control room in case of leaks or exposure during maintenance. Diaphragm transducers liable to contain air-locks. Sensitive diaphragms. Fluid liable to attack manometer liquid. Fluid miscible with manometer liquid.

The second type of error is illustrated by an example. Suppose
we are measuring flue gas flow by a venturi meter placed 25 m above
the instrument (figure A4.3), and the lines are installed as shown;
the upstream line is near the flue and is at 200 °C, the downstream
line is further away and therefore cooler, at 100 °C. Flue gas
velocities are often quite small. The error in the reading may be
calculated by assuming zero flow as in the previous example.

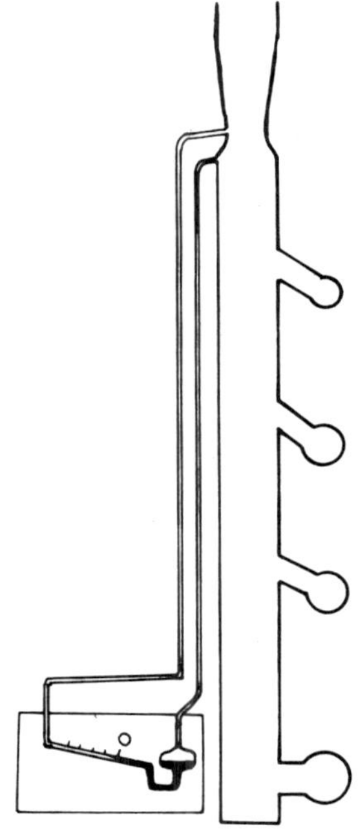

Figure A4·3

In that case the pressure difference at the venturi is negligible;
the pressure at the base of each line is ρgh plus pressure at the
venturi. For the line at 200 °C, assuming the gas in the line is the
same density as air

$$\rho gh = 1.29 \times \frac{273}{473} \times 9.81 \times 25 \ \text{N/m}^2$$

For the other line it is

$$1.29 \times \frac{273}{373} \times 9.81 \times 25 \ \mathrm{N/m^2}$$

Thus

$$\text{meter reading for zero flow} = 232 - 182 \ \mathrm{N/m^2}$$

$$= 5.1 \ \mathrm{mm \ of \ water}$$

On a sloping gauge as in the diagram set at $1 \div 10$ slope, this is a reading of 51 mm.

APPENDIX 5
OPTICAL METHODS

Some features conveniently recorded for measurement by photographs (photogrammetry) are projections of three-dimensional shapes, kinematic effects, waves on water, vibration amplitudes and velocities. With rotary image arrestors, rotating fields can be viewed in real time and analysed.

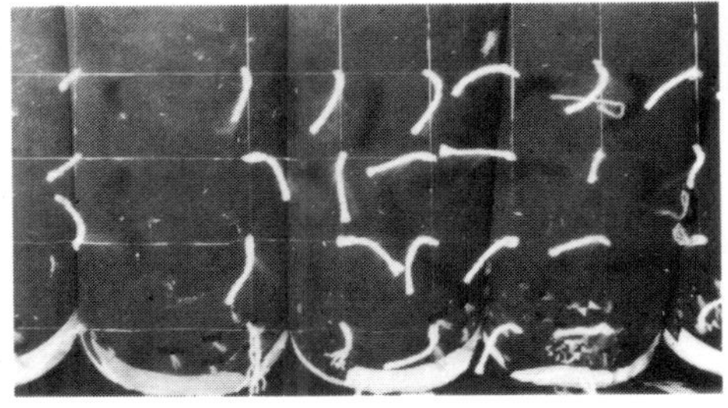

Figure A5·1

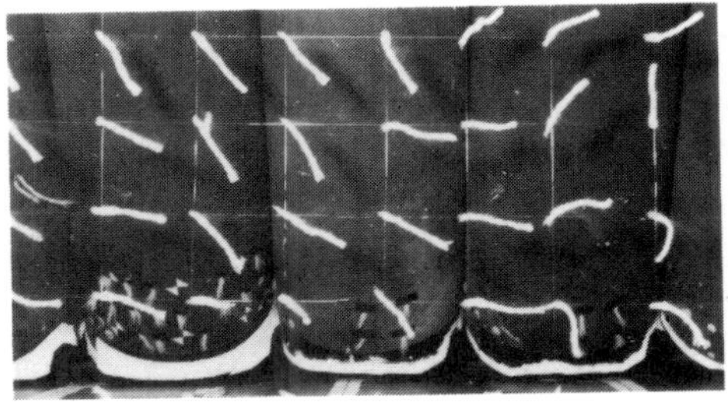

Figure A5·2

Figures A5.1 and A5.2 show examples of a hovercraft skirt, unmodified and with modified edge profile. From such pictures it is possible to measure the net exit area and correlate it with measured flow rates.

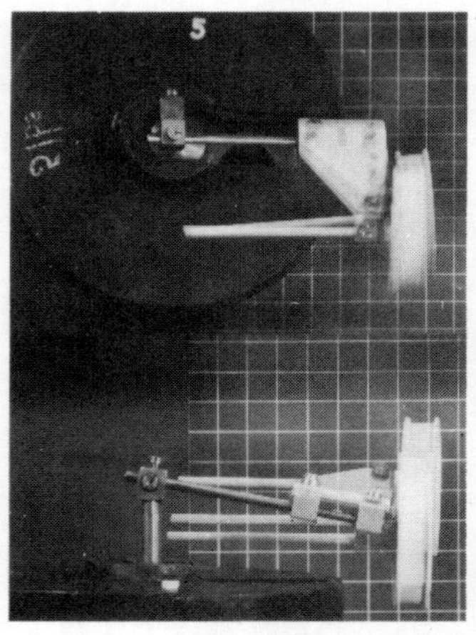 

Figure A5·3   Figure A5·4

   Figure A5.3 shows superposed views of a vehicle suspension model.
The behaviour of such a system is of course capable of analysis by
computer or drawing board but these methods are not only tedious,
they are subject to fallacies. The method shown uses a 45° mirror
to put the plan view into the same picture as the elevation. The
azimuth axle angles of the rear wheel influence the steering response; a heavily laden vehicle tends to swing its rear end more
than when the rear is lightly laden; the suspension geometry is
often arranged to compensate for this. [20]

   In the elevation, the movement of the ground point influences the
choice of spring stiffnesses to optimise comfort without excessive
roll.

   For good results it is best to paint the object matt white.

   An important point is to avoid too many variations in one picture;
then there is no great difficulty in identifying the setting, for
example by the axle ends. Colour film and colour filters can be
used if there is an overlap problem.

   Figure A5.4 shows waves on water with a boat model in a narrow
channel. The picture shows the bow wave height, water level upstream
and alongside, and a drag reading. It is often possible by a little
ingenuity to bring several relevant data into a picture.

   Figures A5.5 and A5.6 show how simply photographs can record
movement. The paper targets are of low mass, stiff enough for low
frequencies and carry a size calibration automatically. A still

112

Figure A5·5   Figure A5·6

picture gives the static position of the wheel when on a vibratory
table in the low or zero position; the moving picture indicates
the unsymmetrical nature of the movement. With stroboscopic
methods the motion could be captured in more detail.

Moving vehicles can be analysed by streak photography. Figure
A5.7 shows an experiment in which a vehicle was driven over a ramp
at various speeds. Lights were attached to the front stub-axle, to
the wing above it and amidships on the bodywork. Note the caption
written into the picture by a torch. Note also a ripple in the
light-streaks. This was a speed reference obtained by pulsing the
lamp current at 10 Hz which was too fast for the lamp filament to

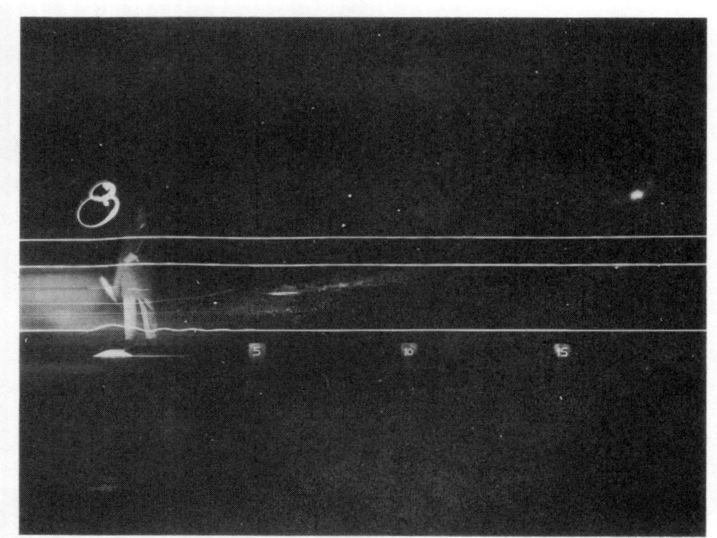

Figure   A5·7

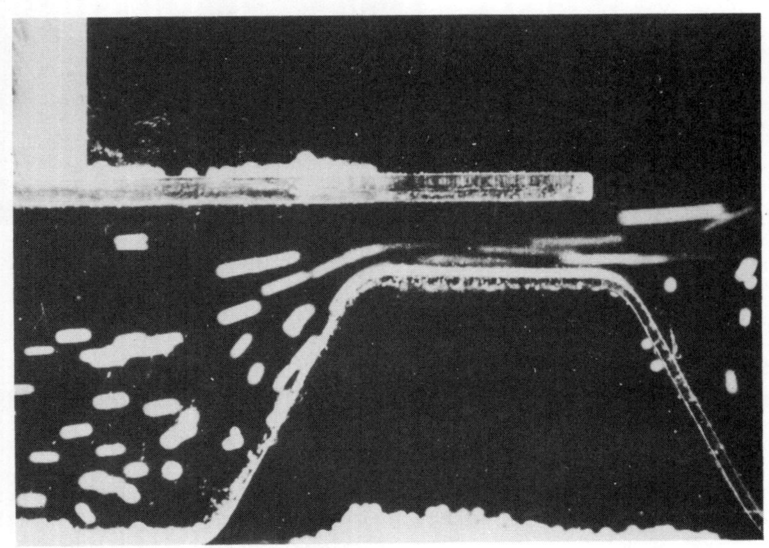

Figure   A5·8

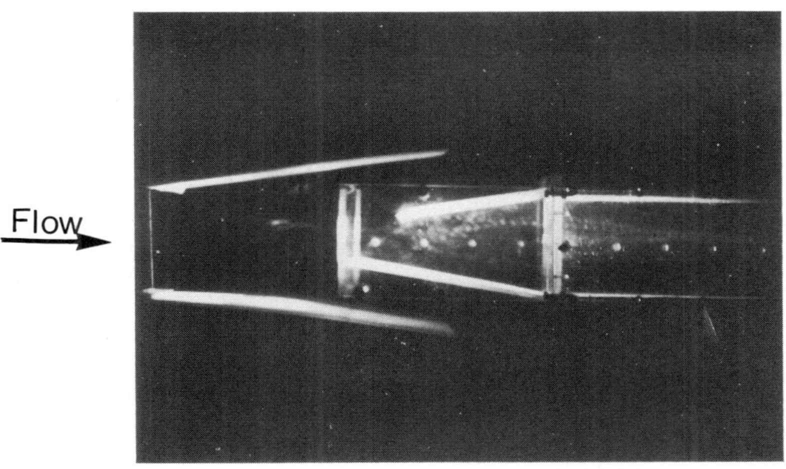

Figure A5·9

respond well, 5 Hz would have been better. These were taken in darkness but this is very inconvenient. Dusk and a dark background should be quite sufficient.

An inexpensive way of recording vibrations uses Christmas tree festoon bulbs, flashed at up to 10 Hz. These are very light and rugged, standing up to flashing and vibration.

Time-lapse photography for speed measurement is well known. Figure A5.8 shows a simple time exposure shot of a water model containing a simulated valve seat, while figure A5.9 shows airflow at $\sim$ 30 m/s visualised with foam granules (see chapter 5 on discussions of validity), photographed through the simple stroboscope shown in figure A5.10.

Figure A5·10

115

Finally the student is again warned about the effects of focal
plane shutters, shown by figure A5.11; the same object, a disc with
four equi-spaced dots photographed with an inter-lens shutter is
shown in figure A5.12. The unequal streaks are illusory, they arise
from the time-lag as the shutter slit passes across the field of
view.

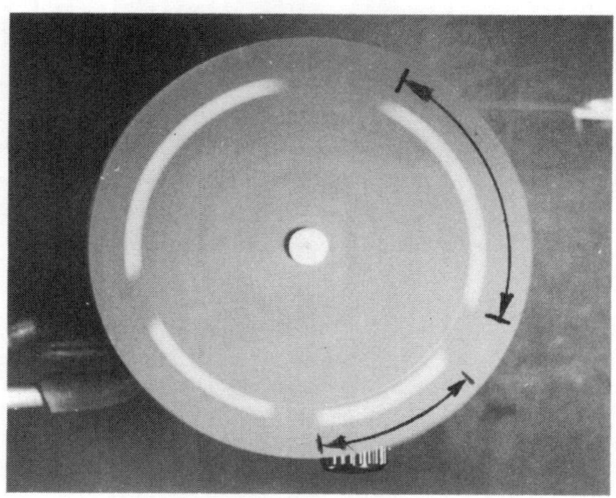

Figure A5·11

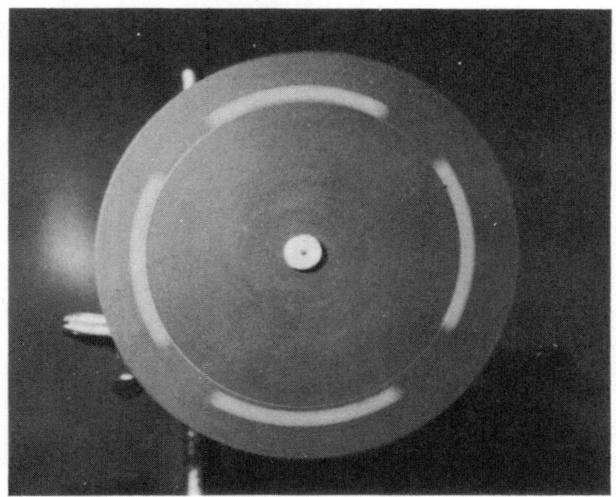

Figure A5·12

Rotation can be arrested by the half-speed prism principle. There
are many references in published literature to researchers using
these devices. The commonest is the dove prism (figure A5.13a) which
refracts before it internally reflects. It is subject to slight
chromatic aberration. A better prism is shown at (b). The base is
silvered, it works entirely by reflection and it is free from chromatic aberration. The roof-top prism (c) may be used in back-reflecting
mode, with a half-silvered mirror to view the result. [21] This type
is easier to build but less convenient to use.

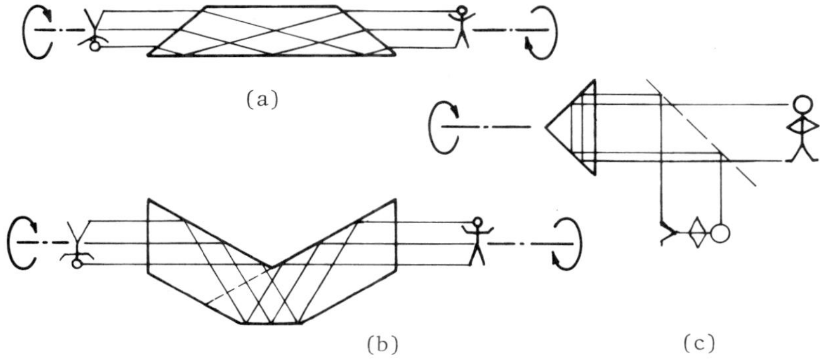

(a)

(b)   (c)

Figure   A5·13

To understand the way a reversing prism arrests rotation requires
some imagination; figure A5.14 should help. The cup on the right is
the object. The upright prism (a) gives an inverted image but no

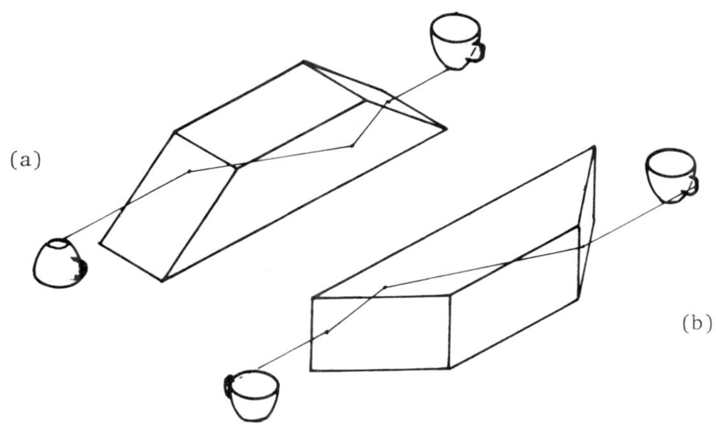

(a)

(b)

Figure   A5·14

117

right-to-left inversion. When the prism has turned 90° (b), the
image is reversed right to left but remains upright; thus it is 180°
different from the first image.

To visualise intermediate positions, imagine the object tilted 20°
anticlockwise, handle upwards. The image (a) will have tilted 20°
handle downwards, since the prism reverses the verticals. Now
imagine *everything* rotated 20° clockwise. The object will be level
again but the image will have rotated further, 40° clockwise.

The argument is confirmed in (b). Again rotate the object on the
right by 20° handle upwards. The image handle must also rotate upwards since there is no change vertically when the prism is on its
side; thus the image has again turned 20° clockwise because the
handle is on the opposite side this time. Rotating the whole clockwise by 20° again restores the object to upright. The image has
turned through 2 × 20°.

These prisms are quite difficult to set up since the axis of the
undeflected ray must lie on the axis of rotation, free from position
error and from angle error.

APPENDIX 6   REFERENCES

1. *International Critical Tables, Volume VI* (McGraw-Hill, Maiden-
    head) 226-7.
2. M. G. W. Dingle, 'Instruments for the Fuel Technologist',
    Ind. Process Heat., 11 (1971) 6-9.
3. G. W. Kaye and T. H. Laby, *Tables of Physical and Chemical
    Constants* (Longman, London, 1966) 40, 67.
4. BS 1041: Part 1: Temperature Measurement.
5. Tempil Division, Big Three Industries Inc., Hamilton Boulevard,
    South Plainfield, N.J. 07080; or Optimum Heat Control Ltd.,
    Station Works, Dedmere Road, Marlow, Bucks.
6. Synthetic and Industrial Finishes Ltd., Imperial Works, Balmoral
    Road, Watford, Herts.
7. Civil Aircraft Accident Report, Comet G-ALYP, 10.1:54 and Comet
    G-ALYY, 8.4.54 (H.M.S.O., 1955).
8. T. E. Taylor, 'Effect of Test Pressure on Fatigue Performance
    on Mild Steel Cylindrical Pressure Vessels containing Nozzles',
    Br. Weld. J., 14 (1967) 461.
9. P. Polak, 'Spontaneous Hot Zone Formation in Oil Flow', *J. mech.
    Engng Sci.*, 13 (1971) 293. See also J. E. Gerrard *et al.*,
    'Viscous Heating in Capillaries, the Adiabatic Case', *Ind. Engng
    Chem. Fundam.*, 4 (1965) 32.
10. J. M. Piau, 'Flow of Dilute Polymer Solutions', International
    Conference on Drag Reduction, Sept. 1974, BHRA Fluid Engineering
    Ltd, Cranfield, Beds.
11. BS 1042: Part 1: 1964 Orifice Plates, Nozzles and Venturi Tubes;
    Part 3: 1965 Guide to the Effects of Departure from the Methods
    in Part 1.
12. American Society of Mechanical Engineers Power Test Code 19.5,
    Sections 2-5.
13. R. C. Pankhurst and D. W. Holder, *Wind Tunnel Technique* (Pitman,
    London, 1968) 193.
14. *Handbook of Chemistry and Physics* (The Chemical Rubber Co.,
    Cleveland, Ohio) E132.
15. Drager Normalair Ltd, Kitty Brewster, Blyth, Northumberland; or
    Dragerwerk A.G., 53/55 Mosslinger Allee, D 24 Lubeck 1, W.
    Germany.
16. Sir William Hawthorne, 'Energy, a Renewed Challenge to Engineers',
    Proc. Instn mech. Engrs, 189 (1975) 446.
17. R. B. Heywood, *Designing against Fatigue* (Chapman & Hall, London,
    1962) section 15.10, 432.
18. S. Timoshenko and J. Goodier, *Theory of Elasticity* (McGraw-Hill,
    Maidenhead, 1970) section 23.
19. ibid., section 118.
20. P. Polak, 'Rear Suspension Geometry Effects', *Automot. Des. Engng,*
    7 (1968) 52.
21. P. Waddell and B. Smart, 'A Non-stroboscopic System for Examining
    High-speed Rotating Objects', *Strain,* 9 (1973) 73.

ANSWERS TO QUESTIONS

CHAPTER 1

1.1 (a) 2000.001 Ω, (b) 0.0025 Ω (2.5 $\mu\Omega$) 1.2 (a) 666.7 Ω,
(b) 1499.997 $\mu\Omega$ 1.3 (a) 1 999 990, (b) 499 990, (c) 1990, (d) 10 Ω,
(e) nil 1.4 (a) 0.0001, (b) 0.0050025, (c) 1.111 Ω, (e) nil
1.5 Removing sample removes competition from surrounding plants;
uprooting disturbs soil (better to cut?); access will influence
some plants by trampling but sampling from edge may not be representative; water content inconsistent? cut only after rain or
dew?

CHAPTER 2

2.1 7.5 mV 2.2 246 °C 2.3 (a) -2 °C, (b) 0, (c) 97 °C 2.4 Yes.
Copper melts at 1064 °C, Constantan at about 1210 °C. 2.5 (a)
18.5 - 0.8 = 17.7 mV, (b) 18.5 - 2.4 = 16.1 mV, (c) 18.5 - 2.4 -
(2.4 - 0.8) = 14.5 mV 2.6 Resistance correction 120/150. Estimate
graph slope, say, 6.25, 56 and 33 μV/°C (a) 1.00 mV, (b) 0.896 mV,
(c) 0.53 mV 2.7 Thermometer zero error or warm water accumulating
in jar. See section 2.6 2.8 (a) 0 °C; immersion beyond mercury
gives no error, (b) 79.24 °C 2.9 (a) 101.5 °C (2° stem error,
-0.5° zero error), (b) 100.75°, see figure 2.5.

CHAPTER 3

3.1 1.2 3.2 0.072, 0.078 3.3 3.51, 3.72 mm 3.4 Calculate bending
moment from deflection. Loads 28.3 N, 27 N

CHAPTER 4

4.1 4.7 N m, 3.8 N m 4.2 5.1 N, 25.5 N, 6.4 N/mm^2, 32 N/mm^2,
679 Hz, 304 Hz 4.3 0.0195 mm, 0.039 mm, 0.195 mm; 0.098 J, 0.34 J,
2.34 J

CHAPTER 5

5.1 0.033 m^3/s, 32.5 kg/s 5.2 24.6 mm water, 0.756 kg/s, 381 mm,
401 mm water 5.3 27 m/s, 5.55 kg/s

CHAPTER 6

6.1 2.36, 2.5, 2.13 t

CHAPTER 7

7.1 37.7 N, 10.37 N, 23.5 N